THE SCIENTIFIC REVOLUTION REVISITED

The Scientific Revolution
Revisited

Mikuláš Teich

http://www.openbookpublishers.com

Digital material and resources associated with this volume are available at http://www.openbookpublishers.com/9781783741229#resources

ISBN Paperback: 978-1-78374-122-9
ISBN Hardback: 978-1-78374-123-6
ISBN Digital (PDF): 978-1-78374-124-3
ISBN Digital ebook (epub): 978-1-78374-125-0
ISBN Digital ebook (mobi): 978-1-78374-126-7
DOI: 10.11647/OBP.0054

Cover image: Giuseppe Arcimboldo, 'The Summmer' (1563), http://commons.wikimedia.org/wiki/File:Giuseppe_Arcimboldo_-_Summer_-_Google_Art_Project.jpg

All paper used by Open Book Publishers is SFI (Sustainable Forestry Initiative) and PEFC (Programme for the Endorsement of Forest Certification Schemes) Certified.

Printed in the United Kingdom and United States by Lightning Source for Open Book Publishers

To the memory of

Alistair Crombie (1915-1996)
Rupert Hall (1920-2009)
Joseph Needham (1900-1995)
Roy Porter (1946-2002)

scholars most learned and friends most loyal

Contents

List of Illustrations

Note on Terminology and Acknowledgements

In a book about the much debated Scientific Revolution, problems unavoidably arise with terminology. They pertain to terms such as science/normal science/modern science, and social/societal, among others. I regret possible ambiguities in their employment despite efforts to be consistent. There is also the question of references. They are given in full but I apologise for inadvertent omissions. This also applies to the bibliography relevant to the debate.

I am indebted to Dr Albert Müller, who read a large part of the early version of the book, and Professor Sir Geoffrey Lloyd, who commented on chapters 1 and 2 in draft. It is a pleasure to pay tribute to discussions with Professors Kurt Bayertz, Herbert Matis, Michael Mitterauer, Dr Deborah Thom and Professor Joachim Whaley. Deep thanks for support are due to Dr Ian Benson, Alison Hennegan, Professor Hans-Jörg Rheinberger and the late Professor William N. Parker. Dr Alessandra Tosi provided invaluable editorial guidance. Ben Fried proofread the manuscript and commented on it most helpfully. Lastly and firstly, my warmest words of gratitude go to my family, above all to Professor Alice Teichova and our daughter Dr Eva Kandler – without their assistance the book would quite literally not have seen the light of day. The responsibility for the published text is mine.

Preface

In 1969, after taking up a Visiting Scholarship at King's College, Cambridge, I was approached by the Department of the History and Philosophy of Science, Cambridge University, to give a public lecture. The subject-matter I chose was 'Three Revolutions: The Scientific, Industrial and Scientific-Technical'. When it was announced in the *University Reporter* (100 (1969-1970), p. 1577), for some reason the Scientific-Technical Revolution metamorphosed into the Scientific-Industrial.

I gave the lecture on 4 May 1970, and in it I attempted to convey that the Three Revolutions were products of, and factors in, historically far-reaching societal transformations, and that the place of science and technology cannot be left out of the societal picture. It was this perspective that led me to return to the subject-matter and address it now in book form.

Apart from underestimating the difficulties of presenting a short account of the issue, other commitments prevented me from focusing solely on the project. When I reached my 90th birthday, it occurred to me that if I was to contribute to the debates regarding these three great movements of thought and action, a viable course would be to produce the work in three separate parts, of which *The Scientific Revolution Revisited* is the first. It turned out to be a thorny journey; the other two parts are in preparation.

Autumn 2014

Introduction

I

This book is about interpreting the Scientific Revolution as a distinctive movement directed towards the exploration of the world of nature and coming into its own in Europe by the end of the seventeenth century. The famed English historian Lord Acton (1834-1902) is said to have advised that problems were more important than periods. If he held this opinion, he ignored that problems are embedded in time and place and do not arise autonomously. The inseparability of problem and period has been amply demonstrated in six collections of essays, examining the 'national context' not only of the Scientific Revolution but also of other great movements of thought and action, which Roy Porter and I initiated and co-edited.[1]

In general terms, one way of encompassing the world we live in is to say that it is made up of society and nature with human beings belonging to both.[2] It is reasonable to connect the beginnings of human cognition of inanimate and animate nature (stones, animals, plants) with the ability to systematically make tools/arms within a framework of a hunting-and-gathering way of life, presently traceable to about 2.5 million years ago. It is also reasonable to perceive in the intentional Neanderthal burial, about 100,000 years ago, the

---

1 Published by Cambridge University Press, the volumes formed part of a sequence of twelve collections of essays which included *The Enlightenment in National Context* (1981), *Revolution in History* (1986), *Romanticism in National Context* (1988), *The Renaissance in National Context* (1991), *The Scientific Revolution in National Context* (1992), *The Reformation in National Context* (with Bob Scribner, 1994), and *The Industrial Revolution in National Context: Europe and the USA* (1996).
2 What follows, draws on the 'Introduction', in M. Teich, R. Porter and B. Gustafsson (eds.), *Nature and Society in Historical Context* (Cambridge: Cambridge University Press, 1997).

http://dx.doi.org/10.11647/OBP.0054.08

earliest known expression of overlapping social and individual awareness of a natural phenomenon: death.

While the theme of the interaction between the social, human and natural has a long history, there is scant debate over the links between perceptions of nature and perceptions of society from antiquity to the present. This is crucial, however, not only for understanding the evolution of our knowledge of nature as well as our knowledge of society, but also for gauging the type of truth produced in the process. An inquiry into the relationship between science and society takes us to the heart of the issue highlighted by the late Ernest Gellner, noted social anthropologist and philosopher, when he stated that 'The basic characteristics of our age can be defined simply: effective knowledge of nature does exist, but there is *no* effective knowledge of man and society'.[3]

This assertion, indeed Gellner's essay as a whole, gives the impression of a despondent social scientist's *cri de coeur*, made before he sadly passed away with the text yet to be published. By then, Gellner had undeniably come to believe that social knowledge compared badly with natural knowledge. He particularly reproved Marxism because it

> claimed to possess knowledge of society, continuous with knowledge of nature, and of both kinds – both explanatory and moral-prescriptive. In fact, as in the old religious style, the path to salvation was a corollary of the revelation of the nature of things. Marxism satisfied the craving of Russia's Westernizers for science and that of the Russian populist mystics for righteousness, by promising the latter in terms of, and as fruit of, the former.[4]

It is noteworthy that this critique contrasts with Gellner's position five years before the demise of communism in Central and Eastern Europe, a development which he clearly had not envisaged:

> I am inclined to consider the reports of the death of Marxist faith to be somewhat exaggerated, at least as far as the Soviet Union is concerned. Whether or not people positively believe in the Marxist scheme, no coherent, well-articulated rival pattern has emerged, West or East, and as people must need to think against some kind of grid, even (or perhaps especially) those who do not accept the Marxist theory of history tend to lean upon its ideas when they wish to say what they do positively believe.[5]

---

3   E. Gellner, 'Knowledge of Nature and Society', cited in ibid., p. 9.
4   Ibid., p. 13.
5   E. Gellner, 'Along the Historical Highway', *The Times Literary Supplement*, 16 March 1984.

This was in line with what John Hicks noted a year after receiving the Nobel Memorial Prize for Economics (in 1972). Venturing to develop a theory of history 'nearer to the kind of thing that was attempted by Marx', he declared:

> What remains an open question is whether it can only be done on a limited scale, for special purposes, or whether it can be done in a larger way, so that the general course of history, at least in some important aspects can be fitted into place. Most of those who take the latter view would use the Marxian categories or some modified version of them; since there is so little in the way of an alternative version that is available, it is not surprising that they should. It does, nevertheless, remain extraordinary that one hundred years after *Das Kapital*, after a century during which there has been enormous developments in social science, so little else should have emerged. Surely, it is possible that Marx was right in his vision of logical processes at work in history, but that we, with much knowledge of fact and social logic which he did not possess, and with another century of experience at our disposal, should conceive of the nature of those processes in a distinctly different way.[6]

'Learning from history' is invoked by politicians at will, but avoided by historians. They could do worse than to heed Hicks's observation regarding Marx's approach to encompassing and deciphering human social evolution. It has not lost its force when it comes to analysing the roots of the contemporary troublesome state of world affairs, fuelled by globalisation.

II

There is no point here in recapitulating what is argued in the book. But, as I have found the strongly-disputed Marxist conception of a period of transition from feudalism to capitalism a useful framework within which to locate the forging of the Scientific Revolution, it may be worthwhile to dwell on it briefly.

According to the Marxist historian Eric Hobsbawm,

> the point from which historians must start, however far from it they may end, is the fundamental and, for them, absolutely central distinction between establishable fact and fiction, between historical statements based on evidence and subject to evidence and those which are not.[7]

---

6 J. Hicks, *A Theory of Economic History* (repr. Oxford: Oxford University Press, 1973), pp. 2-3.
7 E. Hobsbawm, *On History* (London: Weidenfeld & Nicolson, 1997), p. viii.

But what is established fact? Take the categories 'feudalism' and 'capitalism'.[8] There are historians who find them to be of little or no use. There are others who may, curiously, employ both variants in a text: feudalism/'feudalism' and capitalism/'capitalism'. In other words, the categories have the semblance of both fact and fiction. More often than not, the assessment that feudalism and capitalism are not viable historical categories is politically and/or ideologically motivated. This of course is vehemently repudiated on the basis that true historical scholarship does not take sides.

In this connection, Penelope J. Corfield's 'new look at the shape of history, as viewed in the context of long-term-time' comes to our attention. Her interest in this question was triggered by the Marxist historians E. P. Thompson and Christopher Hill (her uncle). Though she clearly disagrees with their world-view, she hardly engages with their work. Criticising the old 'inevitable Marxist stages', she finds that gradually

> over time, historical concepts become overstretched and, as that happens, lose meaning. And 'capitalism'/'communism' as stages in history, along with 'modernity', and all their hybrid variants, have now lost their clarity as ways of shaping history. To reiterate, therefore, the processes that these words attempt to capture certainly need examination – but the analysis cannot be done well if the historical labels acquire afterlives of their own which bear decreasingly adequate reference to the phenomena under discussion.[9]

Corfield's model of making sense of the past is that '*the shape of history has three dimensions and one direction*'. The three dimensions, she argues, are 'persistence/ microchange/radical discontinuity'.[10] While her long-view approach is to be welcomed, her formula gives the impression of being too general to be of concrete value in casting light on, say, the Scientific Revolution.

III

The Scientific Revolution in National Context (1992) illustrated that no nation produced it single-handed. So in what sense was the Scientific Revolution a

---

8   'Once you accept that feudalism existed, and capitalism does, there's a big academic debate about what caused the collapse of feudalism and the rise of capitalism. Shakespeare managed to get to the essence of it without having knowledge of the terms'. Paul Mason (economics editor of Channel 4 News), 'What Shakespeare Taught Me about Marxism and the Modern World', *The Guardian*, 3 November, 2013.

9   P. J. Corfield, *Time and the Shape of History* (New Haven, CT and London: Yale University Press, 2007), pp. ix, 182-83.

10   Ibid., p. 248.

distinctive movement? In the sense that in Europe it had brought into being 'normal science' as *the* mode of pursuing natural knowledge – universally adopted in time and still adhered to at present. Thus 'when an Indian scientist changes places with an Italian or an Argentinian with an Austrian, no conceptual problems are posed. Nobel Prizes symbolise the unity of science to-day'.[11]

In Europe diverse social, economic, political and ideological conditions brought together the historically-evolved ways of knowing nature and produced the Scientific Revolution. These conditions included procedures, such as classification, systematisation, theorising, experimentation, quantification – apart from observation and experience, practised from the dawn of human history. Still, the social context of this transformation of the study of nature into normal science – institutionalised over time and in certain places – may be understood in terms of the passage from feudalism to capitalism. It was a long-drawn-out process of which the Renaissance, the Reformation and the Enlightenment, along with the Scientific Revolution, form 'historically demarcated sequences'.[12] By the eighteenth century, normal science had arrived in latecoming countries, such as Sweden and Bohemia.

Outside Europe the assimilation of normal science had taken place under different historical circumstances. Indeed, we may witness that it still takes place today as part of a fierce global interchange. Existing Stone Age human groups come into contact with latest scientific technology – ancestrally descended from the Scientific Revolution – and eventually they acquire the skills to use electric saws, mobiles, etc., without having passed through the historical learning process experienced by European and non-European peoples under the impact of early capitalist expansion.

The adaptation to tangible contemporary scientific-technical advances by 'primitives' testifies to lasting legacy of the fundamental transformation of the mode of pursuing natural knowledge, both theoretical and practical, between the middle of the sixteenth and the close of the seventeenth centuries. The much maligned Scientific Revolution remains a useful beast of historical burden.[13]

---

11 Introduction in Porter and Teich (eds.), *The Scientific Revolution in National Context*, p. 1.
12 D. S. Landes, *The Unbound Prometheus: Technological Change and Industrial Development in Western Europe from 1750 to the Present* (Cambridge: Cambridge University Press, 1969), p. 1.
13 Introduction in Porter and Teich (eds.), *Scientific Revolution*, p. 2.

1. From Pre-classical to Classical Pursuits

The theme

In the main, historians and philosophers of science have come to differentiate between the *Scientific Revolution* and *scientific revolutions*. The former term generally refers to the great movement of thought and action associated with the theoretical and practical pursuits of Nicolaus Copernicus (1473-1543), Galileo Galilei (1564-1642), Johannes Kepler (1571-1631) and Isaac Newton (1642-1727), which transformed astronomy and mechanics in the sixteenth and seventeenth centuries. First, the Earth-centred system based on Ptolemy's (c. 100-170) celestial geometry was replaced by the heliocentric system in which the Earth and the other then-known planets (Mercury, Venus, Mars, Jupiter and Saturn) revolved around the Sun. Second, laws governing the motion of celestial as well terrestrial bodies were formulated based on the theory of universal gravitation.

The origins of the interpretation of these changes in astronomy and mechanics, made between Copernicus and Newton, as revolutionary are to be found in the eighteenth century.[1] Offering an essentially intellectual

---

[1] I. B. Cohen, 'The Eighteenth-Century Origins of the Concept of Scientific Revolution', *Journal of the History of Ideas*, 37 (1976), 257-88. See also idem, *The Revolution in Science* (Cambridge, MA: Belknap Press, 1985). But Robert Boyle (1627-1691) employed the term 'revolution' to describe the transformation in intellectual life he experienced in the middle of the century. See M. C. Jacob, 'The Truth of Newton's Science and the Truth of Science's History: Heroic Science at its Eighteenth-Century Formulation', in M. J. Osler (ed.), *Rethinking the Scientific Revolution* (Cambridge: Cambridge University Press, 2000). For an instructive account of how writers from Bacon to Voltaire discussed the origins of modern science, see A. C. Crombie, 'Historians and the Scientific Revolution', *Physis:*

http://dx.doi.org/10.11647/OBP.0054.01

treatment of it, Alexander Koyré is credited with having coined the concept of the Scientific Revolution in the 1930s.[2] Since then much has been written about the periodisation, nature and cause(s) of the Scientific Revolution.[3] Broadly, two seemingly incompatible approaches have been employed. The 'internalist' perspective, greatly indebted to Koyré, identified the Scientific Revolution as a societally-disembodied and supremely intellectual phenomenon. The alternate approach, greatly influenced by Marxist ideas, focused on social, political, economic, technical and other 'external' factors to clarify the emergence of the Scientific Revolution.

Since Copernicus's seminal *De revolutionibus orbium coelestium* was published in 1543 and Newton's no less influential synthesis *Philosophiae naturalis principia mathematica* appeared in 1687, some have been perplexed that a phase in scientific history can be called 'revolutionary' when it lasted around 150 years. Others have dwelt on the fact that the protagonists in the transformation of astronomy and mechanics – deemed to be revolutionary – did not fully divest themselves of traditional ancient and medieval approaches and ideas. This connects with the issue of how to view later scientific breakthroughs associated, say, with Antoine-Laurent Lavoisier (1743-1794),

Rivista Internazionale di Storia della Scienza, 11 (1969), 167-80.

2 A. Koyré, *Études galiléennes* (Paris: Hermann, 1939-1940), pp. 6-7.

3 For latter-day discussions of 'the state-of-the-art', see I. Hacking (ed.), *Scientific Revolutions* (Oxford: Oxford University Press, 1981); A. Rupert Hall, *The Revolution in Science, 1500-1750* (London and New York: Longman, 1983), R. Porter, 'The Scientific Revolution: A Spoke in The Wheel?', in R. Porter and M. Teich (eds.), *Revolution in History* (Cambridge: Cambridge University Press, 1986), pp. 290-316; D. C. Lindberg and R. S. Westman (eds.), *Reappraisals of the Scientific Revolution* (Cambridge: Cambridge University Press, 1990); R. Porter and M. Teich (eds.), *The Scientific Revolution in National Context* (Cambridge: Cambridge University Press, 1992); J. V. Field and Frank A. J. L. James (eds. and intr.), *Renaissance and Revolution: Humanists, Scholars, Craftsmen and Natural Philosophers in Early Modern Europe* (Cambridge: Cambridge University Press, 1993); A. Cunningham and P. Williams, 'De-centring the 'Big Picture': The Origins of Modern Science and The Modern Origins of Science', *The British Journal for the History of Science*, Vol. 26/4 (1993), 407-32; H. F. Cohen, *The Scientific Revolution: A Historiographical Inquiry* (Chicago, IL and London: University of Chicago Press, 1994); J. Henry, *The Scientific Revolution and the Origins of Modern Science* (Basingstoke: Macmillan, 1997, 3rd ed. Basingstoke: Palgrave Macmillan, 2008); S. Shapin, *The Scientific Revolution* (Chicago, IL and London: University of Chicago Press, 1998); M. Teich, 'Revolution, wissenschaftliche', in H. J. Sandkühler (ed.), *Enzyklopädie Philosophie*, Vol. 2: O-Z (Hamburg: Meiner, 1999), pp. 1394-97; M. J. Osler (ed.), *Rethinking the Scientific Revolution*; J. P. Dear, *Revolutionizing the Sciences: European Knowledge and its Ambitions, 1520-1700* (Basingstoke: Palgrave, 2001); P. J. Bowler and I. Rhys Morus, *Making Modern Science A Historical Survey* (Chicago, IL and London: University of Chicago Press, 2005), pp. 23-53. P. Fara, *Science: A Four Thousand Year History* (Oxford: Oxford University Press, 2009); David Knight's *Voyaging in Strange Seas: The Great Revolution in Science* (New Haven, CT and London: Yale University Press, 2014) appeared just before this book went to press.

Charles Darwin (1809-1882) or Albert Einstein (1879-1955). Are the novelties of Lavoisier's oxygen theory of combustion, Darwin's theory of evolution or Einstein's linking of space and time comparable in revolutionary terms with the Scientific Revolution? If they qualify as 'scientific revolutions', is the Scientific Revolution then first in time among equals?

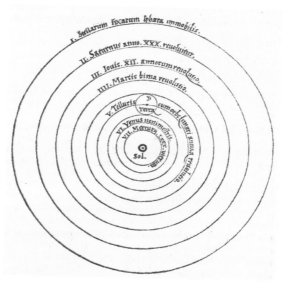

Fig. 1 Image of heliocentric model from Nicolaus Copernicus'
De revolutionibus orbium coelestium (c. 1543).

Kuhn's paradigms and normal science

A determined attempt to address the general question of how scientific revolutions emerge, and how they are identified, has been made by Thomas S. Kuhn in his highly influential *The Structure of Scientific Revolutions*, which first appeared in 1962 and was enlarged in 1970, containing a 'Postscript-1969'. Setting out to portray scientific development (as a succession of tradition-bound periods punctuated by non-cumulative breaks),[4] Kuhn's approach centres on the utilisation of three notions: paradigm, scientific community and normal science. He treats them as mutually connected categories.

For the reader, the grand problem is the truly protean notion of 'paradigm'. After being told that the term he had used in at least 22 different ways, Kuhn

---

4   T. S. Kuhn, *The Structure of Scientific Revolutions*, 2nd revised ed. (Chicago, IL: University of Chicago Press, 1970), p. 208.

admitted: 'My original text leaves no more obscure or important question'.[5] As a consequence, Kuhn preferred to equate a paradigm with 'a theory or set of theories' shared by a scientific community. The question of whether a scientific community's common research activities, designated by Kuhn as 'normal science', determine a paradigm or whether it is sharing a paradigm that defines a scientific community was answered by him as follows: 'Scientific communities can and should be isolated without prior recourse to paradigms; the latter can then be discovered by scrutinising the behaviour of a given community's members'.[6]

To put it succinctly, Kuhn conceives of scientific revolutions as transitions to new paradigms. The motor of this process is not testing, verification or falsification of a paradigm but the scientific community's gradual realisation of a current paradigm's inadequacy. That is, while engaged in normal science, the scientific community finds the paradigm's cognitive utility wanting when confronted with riddles or anomalies which it does not encompass. The response to such a crisis is the emergence of a new paradigm that brings about small as well as large revolutions whereby 'some revolutions affect only the members of a professional subspecialty, and [...] for such groups even the discovery of a new and unexpected phenomenon may be revolutionary'.[7]

The intellectual impact of Kuhn's historical scheme of scientific revolutions was wide-ranging and stimulated much debate during the late 1960s and early 1970s, but it began to wane afterwards. For one thing, on reflection, not only the notion of paradigm but also those of scientific community and normal science appeared to be vague. Take Kuhn's notion of normal science and its association with three classes of problems: determination of fact, matching of facts with theory and articulation of theory. Useful as the concept of normal science is, there is more to it than these three categories, into one of which, Kuhn maintains, 'the overwhelming majority of the problems undertaken by even the very best scientists usually fall'.[8]

Everything has a history and so does normal science. It evolved and materialised first in classical antiquity as *peri physeos historia* (inquiry concerning nature) with entwined elements of scientific methodology, such as observation, classification, systematisation and theorising. By the

---

5   Ibid., p. 181.
6   Ibid., p. 176.
7   Ibid., p. 49.
8   Ibid., p. 34.

seventeenth century in Europe, these practices, extended by systematic experimentation and quantification, were bringing forth generalisations in the form of God-given laws of nature. Moreover, institutionally shored up by newly-founded scientific organisations and journals, these pursuits paved the way for science to operate as a collaborative body. That is, an integral aspect of these developments was the institutionalisation of scientific activities through scientific societies (academies) and journals in Italy, Germany, England and France. Focusing attention on these historical aspects of normal science, we recognise that essentially they still shape its fabric today.

Neither the duration of the coming of normal science into its own nor the blurred line that separates the old from the new in Copernicus's or Newton's thought is the problem.[9] It is the coming into existence of a methodologically-consolidated, institutionally-sustained mode of 'inquiry concerning nature', that distinguishes the investigations into natural phenomena made during the sixteenth and seventeenth centuries from those of previous centuries, and which lies at the heart of the Scientific Revolution.

What the Scientific Revolution arrived at was the eventual institution of science as *the* human activity for the systematic theoretical and practical investigation of nature. In a complex interactive process, intellectual curiosity and social needs were involved and intertwined; and it is not easy to disentangle the 'pure' and 'applied' impulses and motives which advanced the Scientific Revolution. Historically, perhaps, the most significant achievement of the Scientific Revolution was the establishment of science as an individual and socially-organised activity for the purpose of creating an endless chain of approximate, albeit self-correcting, knowledge of nature – a veritable extension of the human physical and physiological means to understand, interpret and change nature.[10]

---

9   K. Bayertz, 'Über Begriff und Problem der wissenschaftlichen Revolution', in his (ed.), *Wissenschaftsgeschichte und wissenschaftliche Revolution* (Hürth-Efferen: Pahl-Rugenstein, 1981), pp. 11-28. Take William Harvey's discovery of the circulation of the blood (1618-1628). It was a product of *both* Aristotelian thinking (in which the idea of the circle plays a major role) and non-Aristotelian quantitative reasoning. See W. Pagel, *William Harvey's Biological Ideas: Selected Aspects and Historical Background* (Basel and New York: Karger, 1967), pp. 73f., J. J. Bylebyl, 'Nutrition, Quantification and Circulation', *Bulletin of the History of Medicine*, 51 (1977), 369-85; A. Cunningham, 'William Harvey and the Discovery of the Circulation of the Blood', in R. Porter (ed. and intr.), *Man Masters Nature: 25 Centuries of Science* (London: BBC Books, 1987), pp. 65-76.

10 J. D. Bernal, *The Extension of Man: A History of Physics Before 1900* (London: Weidenfeld and Nicholson, 1972), pp. 16f.

Empirical knowledge

Relevant to the historical understanding of the Scientific Revolution is the need to distinguish between empirical and scientific knowledge of nature, and to be aware of their historical relations. Broadly considered, empirical knowledge of nature derives from human activity based on observation and experience. Whereas scientific knowledge derives, as indicated, from historically-evolved and interlocked characteristic procedures of investigating nature, including observation.

Observation is an activity not specific to humans. The human perceptual experience of nature, attained through observation, differs qualitatively from that of non-human animals in that it entails mental, verbal, manipulatory and societal dimensions which are hard to disentangle. According to the 'food-sharing hypothesis' propounded by the anthropologist Glyn Isaac, 'the collective acquisition of food, postponement of consumption, transport and the communal consumption at a home base or central place' constituted a major stage in human evolution, assisting 'the development of language, social reciprocity and the intellect'.[11]

It is believed that early humans embarked on producing tools and weapons about 2.5 million years ago. These activities, in combination with meat-hunting and plant-gathering, the use of fire and ability to make and control it, stand at the very beginnings of empirical knowledge of nature. Take the making of stone tools: it involved finding out about the relative hardness and cleavability of stones by trial and error. The underlying dialectic between doing and learning has been pinpointed by the anthropologist Nicholas Toth, who spent many years experimenting with techniques for making stone tools, as follows: 'Toolmaking requires a coordination of significant motor and cognitive skills'.[12]

This applies even more markedly to the manipulative prowess of the modern humans (*Homo sapiens*) who created Palaeolithic art, traceable in the Blombos cave in South Africa to about 75,000 years ago, and in the Chauvet cave in France to about 30,000 years ago. Comparable in age are the Sulawesi cave paintings in Indonesia, pointing to African origins of figurative art before *Homo sapiens* spread across the globe. Explanations and interpretations abound, examining, for example, whether mural pictures of

---

11 G. L. Isaac, 'Aspects of Human Evolution', in D. S. Bendall (ed.), *Evolution from Molecules to Men* (Cambridge: Cambridge University Press, 1983), pp. 532-35.
12 Quoted by R. E. Leakey, *The Origin of Humankind* (London: Basic Books, 1994), p. 38.

animals with arrows in them should be looked upon as a form of hunting magic. Be that as it may, the position of the arrows in the heart region indicates the hunters' familiarity with the (anatomical-physiological) locus where the animal could be mortally wounded. Representations of women with pronounced female sexual attributes (breasts, buttocks, pubic triangle) are evidence that prehistoric humans attached particular importance to fertility and sexual matters. Human interest in reproduction and sexual activity has a prehistoric past.

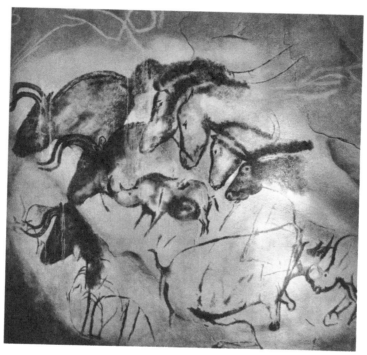

Fig. 2 Palaeolithic painting, Chauvet-Pont-d'Arc Cave
(southern France), c. 32,000-30,000 BP.

It is accepted that the extinct *Homo neanderthalensis* – the evolutionary relations between him and the surviving *Homo sapiens* are still debated – was burying his dead about 100,000 years ago. As previously mentioned, the Neanderthal burials are regarded as the earliest expressions of human awareness of the natural phenomenon of death. With them originates not only the history of human perception of the relation and distinction between life and non-life, but also that of time.

Perception of time and space: early impulses

'Time is a word', we read in an authoritative encyclopaedia of astronomy and astrophysics, 'that eludes definition until it is given some practical application'.[13] The quandary of envisaging time has been reflected in the dichotomy between linear and circular visions of time, depicted vividly by the palaeontologist J. S. Gould as 'time's arrow' and 'time's cycle' respectively. Gould holds that 'time's arrow' – encapsulating the unidirectionality of events – 'is the primary metaphor of biblical history'.[14] Doubtless the lineage of time's cycle is more ancient – it goes back to the hunter-gatherers' observation of recurrent events, such as heavenly cycles, annual seasons or female menstruations.

As to the perception of time's 'twin' – space – it assumes tangible form in terrestrial measurement, in the wake of the growth of permanent agricultural settlements. Heralding the Neolithic Age, agriculture based on the cultivation of soil and the manipulation of plants and animals arrived in parts of Western Asia about 10,000 years ago.[15] It brought about a shift from hunting and gathering to production and storage of food hinged on irrigation and drainage – as in the river valleys of the Nile and the Tigris and Euphrates. The establishment of sedentary life was accompanied by empirically-attained technical developments embodied in a host of arts and crafts, such as pottery, spinning and weaving, dyeing, metal working, house and boat building and others. All these developments contributed to the growth of specialised material production, including that of food. The distribution of products, as well as political, military and religious activities, came under institutional, palace or temple control, administered by officials variously described as 'scribes', 'clerks', 'bureaucrats' – the literate minority of society. Thus the basis was laid for the establishment of socially stratified and centrally governed polities, as encountered in Ancient Egypt and Mesopotamia.[16]

---

13 W. J. H. Andrewes, 'Time and Clocks', in S. P. Maran (ed.), *The Astronomy and Astrophysics Encyclopaedia* (New York: John Wiley & Sons, 1991), p. 929. It is argued that 'on the cosmic (though not human) scale, size is visible in a way that age is not'. See B. Dainton, 'Past, What Past?', *The Times Literary Supplement*, 8 January 2010.

14 S. J. Gould, *Time's Arrow, Time's Cycle: Myth and Metaphor in the Discovery of Geological Time* (Harmondsworth: Penguin, 1990), p. 11.

15 S. Jones, R. Martin and D. Pilbeam (eds.), *The Cambridge Encyclopedia of Human Evolution* (Cambridge: Cambridge University Press, 1992), p. 378.

16 In a study of Ancient Mesopotamia, Susan Pollock shares the view of critics of the 'overarching' notion of the temple economy'. See S. Pollock, *Ancient Mesopotamia The Eden that Never Was* (Cambridge: Cambridge University Press, 1999), p. 119. But even she writes in the concluding chapter: 'Temples, which have been identified as far back as the Ubaid period, are one of the most obvious testimonials to the central place of

Apart from Western Asia where the cultivation of wheat and barley began, other sites of origin for agriculture are recognised. China for rice and millet, for example, or Central America and the northern Andes for maize. Agriculture as a means of supplying the human demand for food turned out to be a worldwide activity. Even today the majority of the world population lives off the land. Because of its unprecedented impact on world history – comparable with the Industrial Revolution – the changeover to economies sustained by agriculture during Neolithic times deserves to be called *the* Agricultural Revolution.[17]

River valley civilisations and knowledge of the natural world

The Neolithic agricultural and craft activities, developed in contact with living and non-living things, broadened the empirical knowledge of diverse natural materials, as well as natural and artificial processes, enormously. The concurrent inventions related to the state, commercial and communication needs of the river valley civilisations – such as measures and weights, numerical symbols and arithmetic, writing and the alphabet – were historically of incalculable import.[18]

The advent of agriculture activated astronomical observations, and with them brought forth the measurement of time as realised in the construction of the calendar. Thus in Ancient Egypt, from about 3000 BC, the length of the

religion within Mesopotamian societies. Yet they were also economic and political institutions; any attempt to apply to them our contemporary notions of the separation of religion, politics, and economy forces us to recognise that our concepts are products of a particular history and culture rather than eternal verities'. Ibid., p. 221.

17 See J. Vandermeer, 'The Agroecosystem: The Modern Vision Crisis, The Alternative Evolving', in R. Singh, C. B. Krimbas, D. B. Paul and J. Beatty (eds.), *Thinking about Evolution, Historical, Philosophical and Political Perspectives*, Vol. 2 (Cambridge: Cambridge University Press, 2001), p. 480. Regarding the juxtaposition of the (disputed) Neolithic Revolution and the Industrial Revolution, see C. M. Cipolla, 'Introduction', in his (ed.), *The Fontana Economic History of Europe: The Industrial Revolution* (London and Glasgow: Collins/Fontana Books, 1973), pp. 7-8. See also, by the same author, *The Economic History of World Population* (Harmondsworth: Penguin, 1962), Ch. 1.

18 The case for making writing, developed in Mesopotamia (about 3000 BC), an integral part of the history of weights and measures has been restated by J. Ritter. 'One outcome of this interplay', he writes, 'was of striking importance at the conceptual level – the development of an abstract use of numbers, independent of any metrological system, and the creation of a positional system of base sixty'. See J. Ritter, 'Metrology, Writing and Mathematics in Mesopotamia', *Acta historiae rerum naturalium necnon technicarum. Prague Studies in the History of Science and Technology*, N. S. (1999), 215-41 (p. 239).

year amounting to 365 days was accepted. The number corresponded to the interval between two observed, predictable events that recurred and coincided annually. That is, the agriculturally vital flooding of the Nile and the rising of the brightest star in the sky (known today as Sirius) after its period of invisibility, just before sunrise in July. The Egyptian year became the basis for calendar computation and reform. It was largely this achievement, together with the recognition of the influence of solar and stellar observations on the alignment of those truly towering works of engineering – the pyramids – that made the fame of pre-Hellenistic Egyptian astronomy.

The emphasis on the agricultural context of ancient astronomy should not obscure other factors at play. Certainly a mixture of religion, astrology and politics was a major stimulant for Babylonian solar, lunar and planetary observations. Take the observations of periodical appearance and disappearance of the planet Venus – identified with the goddess Ishtar – extending over two decades (c. 1582-1562 BC). They were copied and referred to for centuries. The observed phenomena were taken to furnish positive or negative omens affecting the future of the ruler (wars), the community (harvests) and the individual (fertility). Historically noteworthy is the intertwining of astronomy and astrology that went into the construction of the equal-sign zodiac. That is, the circle or belt of star clusters through which the sun was thought to move annually. On the one hand, its division into twelve 'signs' of thirty degrees, named after important star groups, amounted to the construction of a system of celestial coordinates – a significant event in the history of mathematical astronomy (c. early fifth century BC). On the other hand, the old 'signs' have retained their astrological connotation for predicting a person's future to the present. Because of precession of the equinoxes, it is necessary to differentiate between slowly revolving constellations and 'fixed' zodiacal signs carrying the same names.[19]

Regarding the Babylonian observations, what matters in retrospect is not their accuracy – seemingly overplayed – but that 'there was a social mechanism for making and recording astronomical observations *and for storing and preserving the records'*.[20] The 'astronomical diaries', as the resulting records

---

19 B. L. van der Waerden, 'Basic Ideas and Methods of Babylonian and Greek Astronomy', in A. C. Crombie (ed.), *Scientific Change, Symposium on the History of Science, University of Oxford 9-15 July 1961* (London: Heinemann, 1963), p. 42f; Precession of the equinoxes was recognised by Hipparchus (second century BC); see J. North, *Cosmos: An Illustrated History of Astronomy* (Chicago, IL and London: University of Chicago Press, 2008), pp. 14, 114.

20 J. Evans, *The History and Practice of Ancient Astronomy* (New York and Oxford: Oxford

are called, contain astronomical as well as meteorological, hydrological and other entries. The oldest are datable to the seventh century BC but, in view of the age of the observations of the planet Venus, the practice of recording must be even older.[21] Comparable are Chinese records of celestial phenomena beginning in the fifth century BC. Among several affinities between Babylonian and Chinese astronomy is that observation of celestial phenomena fell under state control. In China this control found expression in the setting up of the Astronomical Bureau, as part and parcel of the completion of the unification of the realm under the Han dynasty (202 BC-AD 220). The following comment elucidates the situation neatly:

> Celestial portents were not merely natural phenomena, but expressions of the will of Heaven communicated to the ruler as admonition. According to the Chinese theory of monarchy, the supreme ruler was the Son of Heaven, and through him the celestial will was to be transmitted as the basis of social order. Though the Chinese Heaven is neither a creator nor a god in the theological sense – later, seen more philosophically, it *was* the cosmos or natural order itself – it provided criteria for moral and political conduct and thus occupied a crucial position in Chinese political ideology. To supervise the heavenly ritual was the ruler's privilege as well as his duty, for it was an essential service which only he could perform on behalf of its subjects.[22]

In the light of what has been said, the vital role of empirical knowledge of the natural world, bound up with observation and experience, for early human existence and its advance is manifest. What has to be cleared up is that there is more to the human perception of nature than observations of natural phenomena *per se*. Historically, the names of stars and constellations furnish a striking example. It may be assumed that not a few go back to prehistoric times, when hunters watching the sky with the naked eye 'recognised' figures described as a lion, bear, etc. These names reflected the hunters' preoccupation and familiarity with the world of animals. To them, as indeed to the peoples of the river civilisations, the natural world appeared to be 'alive'. Natural and artificial processes appeared to be 'living' and ideas

University Press, 1998), p. 16 (italics – JE).
21 Ibid.
22 K. Yabuuti, 'Chinese Astronomy: Development and Limiting Factors', in S. Nakayama and N. Sivin (eds.), *Chinese Science Explorations of an Ancient Tradition* (Cambridge, MA and London: MIT Press, 1973), p. 93. For an original account of the history of Chinese astronomy, see J. Needham, *Science and Civilisation in China* (Cambridge: Cambridge University Press, 1959), Vol. 3, pp. 169ff. For critical remarks, see N. Sivin, 'An Introductory Bibliography of Traditional Chinese Science. Books and Articles in Western Languages', in Nakayama and Sivin (eds.), *Chinese Science*, pp. 298-99.

about 'livingness' were derived from experiences with, and observations of, human as well as animal bodily functions. It is not difficult to see that the beginning or origin of everything was linked to human/cattle procreation through sexual union. Akkadian texts refer to male and female stones and metals. The production of metals by the smith was imagined as something related to child birth. From this animate/biological angle to metal extraction, the alchemical idea of a 'marriage of metals' ensued, crystallising eventually into the basic chemical concept of 'combination'.[23]

In Mesopotamia there was a socio-political side to the observation of the celestial world which, as we know in retrospect, was to contribute to the demarcation of pursuits of natural knowledge from other human activities. It concerned the prehistory of the idea of a law of nature, a prehistory that comes to light, as it were, in the process of drawing an analogy between the earthly state and the cosmic state. For example, in a late Babylonian 'creation' poem the sun-god Marduk is pictured as the giver of law to stars. According to Joseph Needham, the prodigious student of the comparative history of science, the genesis of the 'conception of a celestial lawgiver "legislating" for non-human natural phenomena' may be viewed against the background of the unification and centralisation of southern Babylonia by Hammurabi (fl. 1700 BC).[24]

Concept of nature: *phusis*

Before the idea of laws of nature could materialise, the notion of 'nature' had to take shape. Termed *phusis* in Greek, the word (like its Latin counterpart

---

23 R. J. Forbes, 'Metals and Early Science', *Centaurus*, 3 (1953-1954), 30.
24 J. Needham, *Science and Civilisation in China* (Cambridge: Cambridge University Press, 1956), Vol. 2, p. 533. The conjecture has been heavily criticised. But Descartes, Leibniz, Newton and others returned to the idea that a heavenly legislator (God) enacted the laws of nature underlying the motion of matter. See W. Krohn, 'Zur Geschichte des Gesetzesbegriffs in Naturphilosophie und Naturwissenschaft', in M. Hahn und H.-J. Sandkühler (eds.), *Gesellschaftliche Bewegung und Naturprozess* (Cologne: Pahl-Rugenstein, 1981), pp. 61-70 (p. 68). It has been noted that Descartes, who more or less established the conception of nature as governed by laws to be discovered by those who investigated it, never talked about laws of nature with regard to refraction or optics in general. See F. J. Dijksterhuis, 'Constructive Thinking: A Case for Dioptrics', in L. Roberts, S. Shaffer and P. Dear (eds.), *The Mindful Hand Inquiry and Invention from the Late Renaissance to Early Industrialization* (Amsterdam: Koninklijke Nederlandse Akademie van Wetenschappen, 2007), pp. 63-4. The discovery that the ratio of the sine of the angle of incidence to the sine of the angle of refraction is constant for any material, the 'law of sines', is ascribed to Descartes (1638).

natura) is etymologically connected with the idea of genesis or birth.[25] *Phusis* is traceable, it has long been acknowledged, to speculations in the sixth and fifth centuries BC regarding natural phenomena by so-called Presocratic natural philosophers, who hailed from Ionian cities in Asia Minor. To all intents and purposes, their approach to natural phenomena was free of myths and interventions by personal gods. This is not to say that these 'earth-bound' Greek inquirers into nature, as well as others (including medical writers) who followed them up to Galen (fl. AD 180), were without religious beliefs.

The relative geographical proximity of the Ionian cities to Egypt and Babylon has prompted recurring debates regarding the impact of the ancient Near Eastern civilisations upon the Greek world. Going back to the sixth and fifth centuries BC, the knowledgeable classicist Geoffrey Lloyd confirms that both transmissions and independent developments (writing, numerical notation) took place. Lloyd validates noticeable differences between pre-Greek geometry and astronomy. The Near East possessed knowledge of geometrical truths (e.g. the properties of the 'Pythagorean' right-angled triangle) but not the notion of the proof of geometrical truths, something which did develop in Greece. Whilst Babylonian astronomical practice employed arithmetical procedures with respect to planetary movement, the Greeks turned to geometrical models. As for medicine, Lloyd points out that it 'was one of the chief battlefields on which the attempt to distinguish between the "rational" and the "magical" was fought'. This struggle found expression in the Hippocratic collection of Greek medical texts – the oldest of which belonged to the beginning of fifth century BC – in which magical practices and beliefs come specifically under attack.[26]

What is significant is that no other ancient civilisation evolved a notion of nature equivalent to *phusis*. Multifaceted and disputed as the concept of *phusis* was, it stood effectively for objective, intelligible reality and was thus susceptible to rational inquiry.[27] This was connected to a belief in the

25 See entry 'Nature', in W. F. Bynum, E. J. Brown and R. Porter (eds.), *Dictionary of the History of Science* (London and Basingstoke: Macmillan, 1981), p. 289.

26 G. E. R. Lloyd 'The Debt of Greek Philosophy and Science to the Ancient Near East', in his, *Methods and Problems in Greek Science: Selected Papers* (Cambridge: Cambridge University Press, 1991), pp. 278-98.

27 G. E. R. Lloyd, 'Greek Antiquity: The Invention of Nature', in G. Torrance (ed.), *The Concept of Nature: The Herbert Spencer Lectures* (Oxford: Clarendon Press, 1992), p. 22. Reprinted as 'The Invention of Nature', in Lloyd, *Method and Problems*, p. 432. Aristotle (384-24) in *Physics* seems to be the first to have formulated it clearly: 'Nature is a principle of motion and change... We must therefore see that we understand what motion is; for if it were unknown, nature too would be unknown'. See M. Oster (ed.), *Science in Europe*

orderliness of the cosmos – a word of Greek origin. Etymologically bound up with the notion of military orderliness, cosmos was used to signify 'order'/'ordered whole' and eventually stood for the world or universe as an ordered entity. The rational inquiry into the origin and make-up of cosmos, inaugurated by the Ionian thinkers, paved the way for knowledge in fields such as medicine, mathematics, astronomy and physics that had to wait 1500 years before it began to be superseded. The concrete attainments of Greek natural philosophers were highlighted by the influential classicist Moses Finley as follows:

> The Hippocratic practice of auscultation of the heart, Euclid's *Elements*, Archimedes' discovery of specific gravity, the treatise on conic section by his younger contemporary Apollonius of Perge, Eratosthenes' estimate of the diameter of the earth to within a few hundred miles of the correct figure, Hipparchus' calculation of the precession of the equinoxes, Hero's steam-operated toys...[28]

No less noteworthy than these achievements is the *modus operandi* that produced them. Underlying them was the unprecedented conviction that the natural as well as the social – perceived as ordered – were comprehensible without recourse to the supernatural. While the originality of this position – an enduring legacy of Greek antiquity diagnosed by some as the 'Greek miracle' – has not been questioned, its origin has been the subject of debate. During the last four decades or so research has gone some way to demystifying, as it were, the phenomenon by looking into its societal context. Here it is pertinent to recall Finley's uncompromising statement regarding slavery:

> This was a universal institution among the Greeks, one that touched upon every aspect of their lives without exception. It rested on very fundamental premises, of human inequality, of the limits authority and debasement, of rights and rightlessness.[29]

What concerns us here is the relevance of ancient Greek slave-owning society to the understanding of ancient Greek 'inquiry concerning nature'.

1500-1800: A Primary Sources Reader (Basingstoke: Palgrave, 2002), p. 8.

28 M. I. Finley, *The Ancient Greeks* (Harmondsworth: Penguin, 1977), p. 123. The *floruit* dates (BC) of the named persons are as follows: Euclid (300), Archimedes (250), Apollonius (210), Eratosthenes (250), Hipparchus (135). Hippocrates (425) almost certainly did not author any of the sixty treatises or so ascribed to him. Hero was active in the first century AD (60).

29 Ibid., p. 148.

Slavery and 'inquiry concerning nature' in ancient Greece

Tradition has it that it was the Ionian city Chios where slaves were first bought from the barbarians around 550 BC. This was also the period of the beginning of early Greek natural philosophy, personified by the Milesians Thales (585), Anaximander (555) and Anaximenes (535). The question of the connection between their naturalistic speculations about the ordered cosmos, as well as those of later Ionian thinkers, and the rise of slavery in ancient Greece has remained problematical. They employed notions drawn from legal, social, military and political spheres, such as justice, equality (*isonomia*), war, strife, rule, contract and others. As pointed out by Lloyd, these concepts are used 'by one Presocratic after another to convey different conceptions of how the world as we know it, made up of a variety of different things, is never the less an ordered whole'.[30]

But what we know about the ideas of Presocratics is fragmentary and largely second-hand. Hence their uncertain connotation with regards to the historically developing system of slavery in Greek city-states – within a democracy practised solely by male citizens.

Here, as in other matters, Aristotle proves to be illuminating. If we turn to *Politics*, one of his late writings, he addresses the nature of slavery. On this subject, he generalises that the ruler/master/slave relationship permeates 'every composite thing where a plurality of parts, whether continuous of discrete, is combined to make a single common whole'. Aristotle gives examples of the (inanimate) case of a musical scale ruled by its keynote or the (animate) case of the body governed by the soul 'with the sway of a master'.[31]

Aristotle's position on the relation of soul and body as well as on the cognate, but more general issue of the relation of form and matter – as opposites – is germane to the exploration of the role of dichotomies in the evolution of scientific methodology. While rejecting the separateness of soul and body, form and matter, Aristotle envisaged their union to be founded on the subordination of body to soul and matter to form.

Aristotle's fundamental notions are hierarchically predicated, as in form, the causes of things or the scale of being. It is insufficiently appreciated how much Aristotle's commitment to hierarchy and order owes to his acceptance

---

30 'Greek Cosmologies', in Lloyd, *Methods and Problems*, p. 150.
31 Aristotle, *Politics*, I, ii, 9-11 (Loeb Classical Library, Vol. 21, transl. H. Rackham) (Cambridge, MA: Harvard University Press and London: Heinemann, 1977), pp. 19-21.

of the naturalness of social and human inequality, manifestly incarnated in the opposition of freedom and enslavement. At the time of the Peloponnesian war (431-404 BC), it is estimated that there were between 60,000 and 80,000 slaves in Athens – the total population (men, women and children, free or enslaved) was about 250,000 to 275,000.[32] Such social reality palpably underlies Aristotle's conviction that authority and subordination of all sorts and kinds conditioned the ordered existence and functioning of both *polis* and *phusis*. *Polis*, the inegalitarian Greek city-state, was the subject matter investigated in *Politics*.

Rooted in observation and experience – the age-old means of gaining knowledge of the world – the idea of opposites (not unlike that of similarity and difference) supplied a vantage point for theoretical and practical classification and systematisation. Aristotle recognised in these procedures attributes of scientific methodology – in effect, its history begins with him.

The propensity for resorting to the value-laden opposites of inferiority and superiority in scientific inquiry, and its place against the background of the Greek system of slavery, is highlighted by Lloyd as follows:

> The Greeks did not deploy opposites to legitimate a single particular type of political regime. But over and over again their uses of opposites mirror an essential feature of the social structures of Greek society, namely the fundamental division between rulers and ruled. A perceived hierarchical distinction within pairs of opposites that we might have expected to have been totally value-free is a feature that is made to do explanatory work in a variety of scientific contexts … In Aristotle's view … male is held to be 'naturally' superior to female, the latter said to be a 'natural' deformity. Again the members of the pairs right and left, above and below, front and back, are strongly differentiated as to value. Right, above and front are the principles (*archē*), first of the three dimensions (breadth, length and depth respectively), and then also of the three modes of change in living beings, mainly locomotion, growth and sensation. Moreover, this doctrine provides him with the basis of his explanation of a range of real or assumed anatomical facts (the relative positions of the windpipe and the oesophagus, those of the two kidneys, the function of the diaphragm and the positions of the vena cava and the aorta) and even further afield it is the principle he invokes in his admittedly tentative discussion of the difficult problem of why the heavens revolve in one direction rather than in the other. The point can be extended to what we might have assumed to be the purely neutral mathematical pair, odd and even. They provide the basis for the Greek classification of integers and are thus fundamental to Greek arithmetic.[33]

---

32 Finley, *Ancient Greeks*, pp. 72, 55.
33 'Greek and Chinese Dichotomies Revisited', in G. E. R. Lloyd, *Adversaries and Authorities*

Lloyd also raises the question of the sociopolitical background of a pervasive element in Greek natural philosophy, mathematics and medicine. That is, the preoccupation of searchers with foundations, certainties and proofs of truthful knowledge in various domains of inquiry. The most telling example of this tendency is provided by the axiomatic-deductive manner of demonstrating geometrical truth that Euclid displays in his *Elements*. Lloyd points out that the astronomer and cosmologist Ptolemy and the physician Galen, canonical figures of Hellenistic science, subscribed to the idea that proof *more geometrico* establishes certainty of knowledge. Lloyd suggests that this may have something to do with the way in which participants in hard-hitting debates and confrontations in the political assemblies and law courts of the city-states argued their case. The winning depended crucially on marshalled evidence and proof.

This approach leads Lloyd to throw open to discussion the vexed issue of the place of experimentation in Greek science. It is particularly striking, he writes,

> that on many of the occasions when deliberate and explicit testing procedures are invoked, the aim was not so much to devise an experimental set-up that could be seen to be neutral between antecedently equally balanced alternatives, but rather to provide further supporting argument in favour of a particular theory. It is remarkable that even in what are some of the best prepared and most systematic experiments carried out in Greek antiquity, the quantitative investigations of the amount of refraction between various pairs of media (air to water, air to glass, and water to glass) reported in Ptolemy's *Optics*, the results have clearly been adjusted to suit his general theory, since they all fitted exactly.[34]

Investigations into Ancient and Greek Chinese Science (Cambridge: Cambridge University Press, 1996), pp. 134-35.

34 Cf. G. E. R. Lloyd, 'Democracy, Philosophy and Science in Ancient Greece', in J. Dunn (ed.), *Democracy: The Unfinished Journey, 508 BC to AD 1993* (Oxford: Oxford University Press, 1993), pp. 41-56 (p. 45).

2. Experimentation and Quantification

Medieval world

At the end of the 1970s, sociologists of science and sociologically-orientated historians of science began to pay attention to experimentation. Even if their claim that experimentation had been neglected was overstated, it is true that historical literature is rich neither in works dealing with experimentation nor with systematisation.

Uncertainties persist regarding experimentation in the medieval world before it began to occupy, jointly with quantification, the centre-stage of scientific activities in the seventeenth century.

This has something to do with the course of the discussion regarding the medieval origins of normal science, stimulated by Alistair Crombie in the early 1950s. It became overshadowed by the debate on the structure of scientific revolutions, engendered by Kuhn's seminal essay and lasting from its publication in 1962 to about the mid-1980s.[1]

---

1   For what is probably the earliest public presentation of the ideas developed in the essay, see T. S. Kuhn's paper 'The Function of Dogma in Scientific Research', presented at the Symposium on the History of Science, University of Oxford, 9-15 July 1961. See A. C. Crombie (ed.), *Scientific Change, Symposium on the History of Science, University of Oxford 9-15 July 1961* (London: Heinemann, 1963), pp. 347-69. The paper was commented on by A. Rupert Hall and Michael Polanyi, respectively (pp. 370-80). Curiously, neither Kuhn nor Hall mentioned that the latter had already employed the term 'paradigm' in the paper 'The scholar and the craftsman in the scientific revolution' (Hall uses the lower case). It was presented to a history of science conference, also attended by Kuhn (Madison, 1-11 September 1957). See M. Clagett (ed.), *Critical Problems in the History of Science* (Madison, WI: University of Wisconsin Press, 1959), pp. 3-29 (p. 19). Here Hall famously states that although the roles of the scholar and the craftsman in the Scientific Revolution are complementary ones, the former holds the prime place in its story (p. 21).

http://dx.doi.org/10.11647/OBP.0054.02

Medieval science as such was not a concern of Kuhn's except in the context of his interpretation of scientific revolutions as paradigm shifts. Not surprisingly, to the author of the renowned *The Copernican Revolution* (1957) the emergence of Copernican astronomy represented a classic case of a paradigm change. Kuhn acknowledges, we should note, the role played by 'external' factors in the emergence or transition to new paradigms. Regarding the replacement of Ptolemaic astronomy by Copernican astronomy, Kuhn lists among external factors: calendar reform, medieval criticism of Aristotle, the rise of Renaissance Neoplatonism 'and other significant historical elements besides'. Nevertheless, he chooses not to address them:

> In a mature science – and astronomy had become that in antiquity – external factors like those cited above are principally significant in determining the timing of breakdown, the ease with which it can be recognized, and the area in which, because it is given particular attention, the breakdown first occurs. Though immensely important, issues of that sort are out of bounds for this essay.[2]

By contrast, Crombie was concerned with the medieval origins of modern science which he associated with the use of experiment and mathematics. First he traced them back to the thirteenth century, if not to earlier times. But by 1961, he stated: 'I have been responsible for claims that now seem to me exaggerated'.[3]

Even so, Crombie's approach to medieval theoretical and practical engagement with the natural world is still valuable. As, for example, when he underlines the importance of the study of medieval technical texts for the understanding of the evolution of experiment:

> The technological writings of the Middle Ages are still relatively unexplored, and yet it seems to me that it is there that one must chiefly look for those habits developed by the demands made by the problems themselves for accurate, repeatable results. These are of the essence in practical life where it matters if you are given short measure or the wrong product, are subjected to incompetent surgery, or arrive at an unintended destination. They are also of the essence in experimental science. For the history of science in the whole

2   T. S. Kuhn, *The Structure of Scientific Revolutions*, 2nd revised ed. (Chicago, IL: University of Chicago Press, 1970), p. 69.
3   See Crombie's 'Contribution to Discussion of Part Three: Science and Technology in the Middle Ages', pp. 272-91, in his (ed.), *Scientific Change*, pp. 316-23. For Crombie's erstwhile statements, see his *Augustine to Galileo: The History of Science A.D. 400-1650* (London: Falcon Press, 1952); *Robert Grosseteste and the Origins of Experimental Science, 1100-1700* (Oxford: Clarendon Press, 1953).

medieval and early modern period, the relations between the intellectual habits and methods of theoretical science and of practical technology present a vast field of research that has scarcely been investigated. The history of 'practical mathematics' in the Middle Ages would especially repay systematic study.[4]

Let us look at the approach of another authority in this field of historical research – Edward Grant. On the question of experiments in the Middle Ages, taken up in the book in which, as one reviewer put it, he distilled 'a lifetime of scholarly research', Grant has this to say:

> Occasional experiments had been made, and mathematics had been routinely applied to hypothetical, though rarely real, problems in natural philosophy. In the seventeenth century, the new scientists applied mathematics to real physical problems and added experiments to the analytic and metaphysical techniques of medieval natural philosophers. The developments did not emerge from a vacuum.[5]

No doubt, Crombie would have had agreed. The problem is the difference in the thinking of the two historians on what constitutes the milieu or, in Grant's words, the 'societal environment in the Middle Ages that eventually enabled a scientific revolution to develop in the seventeenth century'. Basically, Grant equates this environment with (1) the translation of Greco-Arabic works on science and natural philosophy into Latin, (2) the formation of the medieval university and (3) the rise of the theologian-natural philosopher.[6]

What emerges from Grant's account is that medieval savants set themselves suppositional problems and sought intellectual solutions to them.

Crombie impressively returned to this problematic in his massive three-volume *Styles of Scientific Thinking in the European Tradition* (1994), where he discusses the apparent predisposition of Western society to experimental investigation of nature in conjunction with the medieval philosophical theology of the Creator as a divine mathematician. According to this way of thinking, God created the world in which all things were ordered by measure, number and weight. God also created man in his image, endowed with senses and reason to unriddle God's thinking. Such a belief, Crombie argues, offered the way not only to the systematic use of observation, experiment and logical

---

4 Crombie (ed.), *Scientific Change*, p. 319.
5 E. Grant, *The Foundations of Modern Science in the Middle Ages* (Cambridge: Cambridge University Press, 1996), p. 202.
6 Ibid., p. 171.

argument to bring nature under control, but also to the improvement of the human condition. But Crombie contends:

> The habit of systematic measurement and its instrumentation by appropriate procedures was characteristically a response not to the theoretical demands of natural philosophy but to the practical demands of the technical arts. Academic natural philosophy put a premium on logical precision and internal coherence; practical life required exact and repeatable measures of the external world as experienced and used. Technical ability to specify the conditions for producing a desired result was an essential need of theoretical science and of practical art alike; a quantified experimental science depended on a dialogue between the two. That could take place at a suitable level of education. The technical innovations which came to quantify many aspects of practical medieval life in the 12th century were being matched from early in the 12th century by an increasing attention of scholars to the practical arts, both within general encyclopedias and in more specialized treatises. Intellectual contact was encouraged at once by the improved education of superior craftsmen and by the enlargement of the technical content of the university curricula especially in the mathematical *quadrivium*.[7]

Crombie found confirmation for this viewpoint when he considered the evolution of quantification of fundamental entities during 1200-1500: time, space and weight.

Quantification of time: mechanical clock

Regarding the measure of time, it is accepted that the spread of the mechanical clock effected a radical change in Europe beginning around 1300. Its operation depended on the ingenious combination of a driving mechanism (falling weight) and a regulating mechanism ('foliot-and-verge' escapement).

The perception of time as a continuum was transformed by slicing it into concrete, identical small-time portions (*minutae*). This was due to the mechanical clock's capability to indicate the time of the day – reckoned from

---

7   A. C. Crombie, *Styles of Scientific Thinking in the European Tradition: The History of Argument and Explanation Especially in the Mathematical and Biomedical Sciences and Arts* (London: Duckworth, 1994), Vol. 1, pp. 416-17. The notion of a 'superior craftsman' originates with Edgar Zilsel's 'superior artisan'. See his 'Sociological Roots of Science' (1942), reprinted in D. Raven et al. (eds.), *Edgar Zilsel The Social Origins of Modern Science* (Dordrecht, Boston, MA and London: Kluwer, 2000), pp. 7-21. The historian T. Inkster differentiates between 'higher artisanal' and 'lower craftsman' knowledge. See his 'Thoughtful Doing and Early Modern Oeconomy', in L. Roberts, S. Schaffer and P. Dear (eds.), *The Mindful Hand Inquiry and Invention from the Late Renaissance to Early Industrialization* (Amsterdam: Koninklijke Nederlandse Akademie van Wetenschappen, 2007), p. 445.

one midnight to the next – split into equal 24 hours, each containing 60 minutes of 60 seconds. It should be added that the weight-driven mechanical clock was not accurate enough for measuring small intervals of time. Nevertheless, in comparison with the contemporary methods of telling day time, by the sundial and astrolabe respectively, the superiority of the mechanical clock as a timer was obvious. Of the two, the fixed sundial was simpler to handle – shadow indicated the sun's progress through the sky. Though portable and usable by day, or night, the astrolabe was a more complicated timekeeping implement, as was the armillary sphere. They were devices for measuring the position of stars; from the obtained values, it was possible, in the thirteenth century, to calculate time reliably to 2-5 minutes.

When and who actually invented the mechanical clock is unknown – the first firm date is 1286.[8] As the eminent historian of technology Donald Cardwell states:

> Its design may have resulted from the speculations of some millwrights who knew about gearing and the problems of uniform motion, and who, moreover, had astonishing insight into mechanical principles.
> All we can suppose is that there must have been many attempts to devise a machine to indicate the position of the sun in its daily journey round the earth, and therefore to tell the time. Certainly many of the first clocks were astronomical ones, some of them of such elaborate design that the positions of the sun, the moon, the other five planets and even the motions of the tide could be displayed.[9]

Undoubtedly we are on firmer ground when we inquire about social conditions that favoured the invention of this crucially novel device for measuring time. The mechanical clock was a product of the need to regulate the timing of religious and burgeoning multifarious urban (civic) activities as well as a factor in achieving this regulation. It is no accident that the clock ostentatiously came to adorn monasteries, churches and town halls.

---

8 By and large, scholars do not accept the claim by Joseph Needham (and his collaborators Wang Ling and D. J. de Solla Price) that the hydro-mechanical escapement of the astronomical clock described by the eminent Chinese scholar and state servant Su Sung (1088) represents an important stage in the development of the mechanical clock, with its verge-and-foliot escapement of late thirteenth-century Europe. 'The Chinese measured time by the continuous flow of water, the Europeans, by the oscillatory movement of a verge-and-foliot. Both techniques used escapements, but these have only the name in common. The Chinese worked intermittently, the European, in discrete but continuous beats'. D. S. Landes, *Revolution in Time* (Cambridge, MA: Harvard University Press, 1985), p. 21.

9 D. Cardwell, *The Fontana History of Technology* (London: Fontana Press, 1994), p. 41.

Fig. 3 The Prague Astronomical Clock (Prague Orloj)
in Old Town Square, Prague, Czech Republic.

Especially during its early phases, clockmaking was professionally intertwined with astronomy. The ingenuity underlying the making of clocks was regarded so highly that Nicole Oresme (c. 1325-1382), the great French medieval savant, was prompted to visualise God the Creator as a clockmaker. Just as man contrived to produce a self-moving clock, so 'did God allow the heavens to move continually according to the proportions of the motive powers to the resistances and according to the established order (of regularity)'.[10] An early instance of invoking the image of God as the heavenly clockmaker!

Quantification of space: compass and cartography

In some ways, the part played by the magnetic needle compass in the history of space measurement was analogous to that of the mechanical clock in the history of time measurement. The compass is described for the first time in

10 Quoted in J. Kaye, *Economy and Nature in the Fourteenth Century: Money, Market Exchange, and the Emergence of Scientific Thought* (Cambridge: Cambridge University Press, 1998), p. 224.

a Chinese text dating to about 1088, whereas in Europe the first reference occurs in Alexander Neck(h)am's (1157-1217) *De naturis rerum* (c. 1200). A letter known as *Epistola de magnete* by Petrus Peregrinus (Pierre de Maricourt) (fl. 1269) contains a summary of the European knowledge of magnetic phenomena in the late Middle Ages. Peregrinus conceives the compass both as an astronomical and a navigational instrument.[11]

He describes magnetic compasses without and with a pivot and scale. They were in use in the Mediterranean, clearing the way for the drawing of the first medieval maps, known as *portolani* (compass-charts). Made by practical men, Crombie states,

> and based on the direct determination of distances and azimuths by using log and compass, they were specifically guides to coastlines. From the earliest extant examples of the *Carte Pisane* (1274), the portolans showed scales of distances. By the 16th century they gave two essential pieces of information for navigation: the route to follow and the angle it must make with the north-south axis as given by a magnetized needle; and the distance to run in the direction thus determined.[12]

Compass-bearing in conjunction with observations of currents and winds, rather than methods of astronomical observation, guided the first Portuguese and Spanish voyages of discoveries, including Columbus's transatlantic crossing, in the late fifteenth century.[13]

Against this it is pointed out that Ptolemy's influential *Geography* had become well-known in Portugal and Spain before Bartholomew Diaz, Vasco da Gama and Christopher Columbus embarked on their voyages and thus played a role therein. The Latin translation of the work appeared in print for the first time in Florence in the early fifteenth century. It contained maps drawn on a gridwork of parallels and meridians located with respect to the positions of celestial bodies. Of Ptolemy's coordinate system, Crombie notes that 'by its emphasis on an accurate linear measure of the arc of the meridian it came to transform quantitative mapping'.[14]

Crombie also accepts that Ptolemy's work played a part in the rediscovery of linear perspective. That is, Ptolemy showed how to draw a map as a projection from a single viewpoint. It is generally acknowledged that the technique of linear perspective was invented and demonstrated by Filippo

---

11 For an English translation, see *The Letter of Peregrinus*, in E. Grant (ed.), *A Source Book in Medieval Science* (Cambridge, MA: Harvard University Press, 1974), pp. 368-76.

12 Crombie, *Styles*, Vol. 1, p. 420.

13 D. Goodman, 'The Scientific Revolution in Spain and Portugal', in R. Porter and M. Teich (eds.), *The Scientific Revolution in National Context* (Cambridge: Cambridge University Press, 1992), p. 166.

14 Crombie, *Styles*, Vol. 1, p. 433.

Brunelleschi (1377-1446), the eminent Florentine architect, between about 1413 and 1425. Whether he was acquainted with Ptolemy's work is unclear. Be that as it may, the latter guided multitalented men such as P. dal Pozzo Toscanelli (1397-1482), L. B. Alberti (1404-1472), L. Ghiberti (1378-1455), Nicolaus of Cusa (1401-1464) in 'their common search for a quantified space and techniques for its measurement in astronomy, cartography, optics and painting alike'.[15]

Quantification of weight: statics and assaying

Historically, the quantification of weight by measurement has empirical origins going back to the invention of the balance with equal arms. This was in use in Egypt and Mesopotamia from 2700 BC. It seems to have taken about two and a half millennia before the balance with unequal arms was invented.[16] Its principle was known to the author of *Problems of Mechanics* who, it is now accepted, was a follower of Aristotle. He was familiar with the use as well with the properties of the lever. Thus, for instance, he asks:

> Why is it when two men carry a weight between them on a plank or something of the kind, they do not feel the pressure equally, unless the weight is midway between them, but the nearer carrier feels it more? Surely it is because in these circumstances the plank becomes a lever, the weight the fulcrum, and the nearer of the two carrying the weight is the object moved, and the other carrier is the mover of the weight.[17]

What we have here is an empirical recognition of the law of the lever as later presented by Archimedes (287-212) in a formal mathematical language. Archimedes's status as a great researcher in pure and applied mathematics was already acknowledged in antiquity. What has remained unclear was his attitude to practice. On the one hand, the biographer Plutarch (fl. 83) refers to Archimedes's disdain for the work of the engineer and for artisanal activities

---

15 Ibid., p. 455. For a valuable contribution on the relationship between mathematics and painting in the late Middle Ages, see J. V. Field 'Mathematics and the Craft of Painting: Piero della Francesca and Perspective', in his and Frank A. J. L. James (eds. and intr.), *Renaissance and Revolution: Humanists, Scholars, Craftsmen and Natural Philosophers in Early Modern Europe* (Cambridge: Cambridge University Press, 1993), pp. 73-95. Field underlines the Euclidian rather than the Ptolemaic impulse. English seamen in the sixteenth century began to employ astronomical observations and mathematical calculations in navigation instead of relying largely on practical experience. See S. Rose, 'Mathematics and the Art of Navigation: The Advance of Scientific Seamanship in Elizabethan England', *Transactions of the Royal Historical Society*, 14 (2004), 175-84.

16 Here, and in what follows, I draw on P. Damerow, J. Renn, S. Rieger and P. Weinig, *Mechanical Knowledge and Pompeian Balances*, Preprint 145 (Berlin: Max-Planck-Institut für Wissenschaftsgeschichte, 2000). See also G. E. R. Lloyd, *Greek Science after Aristotle* (London: Chatto and Windus, 1973), p. 48.

17 See Grant (ed.), *Source Book*, pp. 223-24, n. 22.

in general. On the other hand, his own inventive abilities are praised by authors such as Polybios (fl. 164 BC) and Livy (59 BC-AD 17). Apart from the device known as the Archimedean screw, he was said to have invented powerful contrivances for lifting and moving heavy loads, and the steelyard.

What cannot be disputed is that Archimedes thought deeply about methodological questions. That is, he was concerned about the truth of a theorem deduced geometrically. He believed that the truth of the proof demonstrated by geometry follows more easily from knowledge acquired previously through contemplation of a mechanical problem.[18]

Archimedes conceived of a mechanical problem as belonging to statics. In effect, he brought into being scientific statics and hydrostatics as a branch of mechanics with weight as its foundational category – a branch that originated empirically and was rooted in reality. This is clearly shown in the well-known, albeit apocryphal, story of Archimedes's discovery of the hydrostatic principle named after him. It enabled him to solve the problem posed by King Hiero of Syracuse as to whether the royal crown was made of pure gold.

Archimedes's prestige was so great that the authorship of medieval works on statics was wrongly ascribed to him. In comparison with his original text, however, a discernible shift occurred in the attitude towards practice. To the 'science of weights' (*sciencia de ponderibus*), as medieval statics was called, the theory underlying the mechanics of moving heavy objects was of interest. Thus the Toledan translator Domingo Gundissalvo (fl. 1140), drawing on Arabic authors, has this to say about the science of weights:

> The science of weights considers weights in two ways: either (1) according to the weights themselves that are being measured or according to what is measured with them and by them; and this is an inquiry about the principles of the doctrine on weights. Or (2) it considers them in so far as they are moved or according to the things with which they are moved; and this is an inquiry about the principles of instruments by means of which heavy bodies are lifted and on which they are changed [or carried] from place to place.[19]

Evidence for the rising awareness of worldly affairs in medieval intellectual circles is provided by the unknown author of the pseudo-Archimedean

---

18 Archimedes's thoughts on this are enshrined in *The Method*, an incomplete treatise. They are addressed to Eratosthenes, renowned for his remarkable method of calculating the circumference of the earth. See the translation in T. L. Heath, *The Works of Archimedes* (Cambridge: Cambridge University Press, 1912). Professor Lloyd has commented to me: 'I don't think you have got mechanics in Archimedes quite right. It is not that he tried a mechanical problem first and then turned to a geometrical analysis. The problem is mathematical from the outset. The use of mechanics is limited to the application of the two ideas that a geometrical figure can be thought of as balanced around a fulcrum'.

19 See Grant (ed.), *Source Book*, pp. 75-6.

treatise *De insidentibus in humidum* (c. 1250). He displays distinct familiarity with price-fixing in the market-place and transfers this insight to the solution of hydrostatic problems:

> Since the size of certain bodies cannot be found geometrically because of their irregular shape, and since the price of certain goods is proportional to their sizes, it was necessary to find the ratio of the volumes of bodies by means of their weights in order to fix their definite prices, knowing the volume ratios from the weight ratios.[20]

Propelled by the silver-based economy, the quantification of weight by measurement came into its own in assaying during the Middle Ages. The purpose of assaying was particularly to test for the amount of gold and silver that could be extracted from ores and to find out whether coins and the precious metals used in jewellery were pure.[21]

The assayer essentially reproduced the large-scale smelting operation quantitatively on a small scale. The process involved the recovery of silver and gold, in a stream of air, from lead beads placed in a shallow dish made of bone ash (cupel). The end of the process was signalled by the appearance of a bead of the precious metal in the dish which could then be weighed. The balance's limit of accuracy was about 0.1 milligram.

In a noteworthy characterisation of the assayers' and the refiners' craft, Rupert Hall observes that it was

> a *quantitative* craft; profit arose from successful use of the balance, for margins were small. Here, as in navigation, science and craft came close together; but while the navigator was the astronomers' pupil, the chemist descended from the assayer.[22]

A striking early illustration of science and craft connecting, in the context of assaying, is by provided by an edict of Philip de Valois in 1343.[23] It contains two *caveats*, as it were, to be observed by the assayers. First, the balance is to be accurate – leaning neither to right nor left. Second, the assayers are

---

20 Quoted by O. Pedersen and M. Pihl, *Early Physics and Astronomy: A Historical Introduction* (London: Macdonald and Jane's and New York: American Elsevier Inc., 1974), p. 210.

21 It is of interest that the words 'test' and 'testing' relate to the Latin '*testa*', meaning an earthenware pot or vessel employed in metallurgical operations.

22 A. Rupert Hall, 'Early Modern Technology to 1600', in M. Kranzberg and C. W. Pursell, Jr. (eds.), *Technology in Western Civilization*, Vol. 1 (New York: Oxford University Press, 1967), p. 94.

23 Here I draw heavily on F. Greenaway's 'Contribution to the Discussion of Part Three: Science and Technology in the Middle Ages', in Crombie (ed.), *Scientific Change*, pp. 329-31.

advised to perform a blank test on a sample of the lead to be assayed. The idea was to find out whether it contained silver and, if so, how much. This is not quantitative chemical analysis, but it is a step towards it.[24]

It was in the late Middle Ages that experimental weighing was beginning to interest the learned as a means of acquiring natural knowledge. Among those who realised its importance and wrote about it was that audacious thinker Nicolaus of Cusa, in his *Idiota: De staticis experimentis* (1450). As Crombie points out, Cusa envisaged a programme and proposed experimental procedures for measuring a wide range of properties and for determining by measurement the composition of different materials. Because, according to Cusa,

> By the difference of weights, I thinke wee may more truly come to the secret of things, and that many things may be known by a more probable conjecture.[25]

Cusa himself did not perform experiments. But the idea of comparing the weight of materials, before their starting and after their finishing, was eventually to lead to one of the great scientific generalisations by Lavoisier – the principle of conservation of matter (1789).

Fig. 4 Portrait of Nicolaus of Cusa wearing a cardinal's hat,
in Hartmann Schedel, *Nuremberg Chronicle* (1493).

24 A. J. Ihde, *The Development of Modern Chemistry* (New York: Evanston; London: Harper and Row, 1964), p. 23.
25 Crombie, *Styles*, Vol. 1, pp. 421-22.

Fermentational and metallurgical contexts[26]

That Lavoisier formulated the principle of conservation of matter – weight of products equals the weight of reactants – from observing the chemical changes that underline the natural process of fermentation has been regarded as somewhat puzzling. But it should not be cause for surprise, seeing that historically a good deal of chemical knowledge evolved from the experience and problems of fermentation.

In effect, the formulation of the principle was the fruit of the interaction of experimentation, quantification and the theory of phlogiston, created by the German chemist Georg Ernst Stahl (1659-1734). Stahl's thinking about chemical transformation owed a good deal to the examination and discussion of processes associated with the preparation of fermented drinks and the making of bread. This was certainly one of his major interests, as indicated by his first significant chemical work, *Zymotechnia Fundamentalis*, published in Latin in 1697 and posthumously in German in 1734.

Georg Erneſtus Stahl, onoldo Francus,
Med. Doct. h.t. Prof. Publ. Ord. Hall. –

Fig. 5 Georg Ernst Stahl. Line engraving (1715).

26 See M. Teich, 'Circulation, Transformation, Conservation of Matter and the Balancing of the Biological World in the Eighteenth Century', *Ambix*, 29 (1982), 17-28.

It is noteworthy that one of the driving forces behind the translation of this work was the high expectancy of its economic effect. In the preface the anonymous translator claims, in the spirit of mercantilism, that Germany could save millions on imports if more attention were paid to the ways in which wines, beers and spirits were produced.

There can be no doubt that the close connection between theoretical chemical knowledge and its practical use accorded well with Stahl's views. Indeed, the actual impetus that got him developing the phlogiston theory was his interest in the process of smelting ores. Stahl elaborated a picture of the reduction process revolving around the release and transfer of a subtle material to the ore, postulated to be present in charcoal, that he came to call phlogiston. That is, he explained the reduction of ores as 'phlogistication' and the combustion of metals as 'dephlogistication'.

Moreover, he envisaged that phlogiston embodied the subtle matter of combustibility that *linked* the vegetable, animal and mineral kingdoms. In fact, he visualised a global circulation of phlogiston. Underlying it was the conjecture that the phlogiston of the air was absorbed by plants during their growth, then taken up in the way of vegetable food by animals, and then passed back into air through breathing.

Regarding Lavoisier, what emerges clearly is the initial impetus he received from the prize-winning essay on wine fermentation and the best way of obtaining alcohol by Abbé François Rozier in 1770. This contained the suggestion that common air played a part in the souring of wine, a process that has worried man since he became involved in its preparation. It was this idea that may have provided the first clue leading to Lavoisier's subsequent interest in the aeriform state: the physical and chemical properties of 'elastic fluids' or 'airs', including the phenomenon of heat and the composition of common air and water. Through investigations of these problems Lavoisier eventually arrived at a new conception of acidity, calcination and reduction of metals – that of combustion and respiration based on oxygen – turning on the principle of conservation of matter.

The analogy between respiration and slow combustion yielding carbonic acid and water was recognised in 1784 by Lavoisier in the quantitative work he did on a guinea pig in an ice calorimeter, conducted jointly with Pierre-Simon Laplace (1749-1827). But it was almost a decade later that this analogy, which included the generation of animal heat, was explained in the light of the transformed chemical thinking by Lavoisier in collaboration with Armand Seguin (1767-1835).

Quantification of qualities: motion, change and money

The reference to the world of commerce in a pseudo-Archimedean medieval treatise may surprise. However, during the last quarter of the twentieth century, studies appeared substantiating the thesis that medieval scholars owed more to their monetised societal setting than was previously conceded or even considered. This is brought to light when we examine the medieval approach to local motion as a problem of quantifying a quality against its socioeconomic background.

But before we return to this, it is useful to recall the strict Aristotelian separation of quality and quantity as incommensurable categories. The Aristotelian denial that qualities could be quantified was called into question, during the fourteenth century, by some of the most eminent scholastics then active in the universities of Oxford and Paris – acknowledged centres of medieval scholarship. Underlying their discussion of 'intension and remission of forms and qualities' was the belief that variations in degree of qualities were measurable (*latitudo qualitatum*).

Their work included queries into the quantifiability of divine grace but also into the quantifiability of local motion, or velocity, comprehended as continuous magnitudes. Here, as in other areas of inquiry into natural and social phenomena, medieval scholarship was confronted with Aristotelian generalisations. Regarding the speed of a body in motion, Aristotle argued that it was directly proportional to its weight ('force') and inversely proportional to the resistance of the medium in which it moved ('density'). Accordingly, any force, however small, could move any resistance, however large.

Perceiving the paradox in Aristotle's position, scholars in Oxford associated with Merton College – known as the Merton School or the Oxford Calculators – challenged it.[27] Conceiving variations in velocity as variations in the intensity of a quality mathematically, they arrived at what became known as the 'mean speed theorem'. It proposed that a uniformly accelerated velocity could be measured by its mean speed. The proposal has been hailed as 'probably the most outstanding single medieval contribution to the history of mathematical physics'.[28]

Other historically notable fourteenth-century challenges to Aristotle's approach to the motion of bodies came from Paris. Certainly Jean Buridan

---

27 The Merton School included Thomas Bradwardine (c. 1290-1349), John Dumbleton (died c. 1349), William Heytesbury (fl. 1235), Richard Swineshead (fl. 1340-1355).

28 Grant, *Foundations*, p. 101.

(c. 1300-c. 1350) was critical of Aristotle's notion that the movement of a projectile depended on the propelling action of the air. He found Aristotle's explanation unsatisfactory because it was contradicted by experience:

> The first experience concerns the top (*trocus*) and the smith's mill (i.e. wheel-*mola fabri*) which are moved for a long time and yet do not leave their places. Hence, it is not necessary for the air to follow along to fill up the place of departure over a top of this kind and a smith's mill. So it cannot be said [that the top and the smith's mill are moved by the air] in this manner.
>
> The second experience is this: A lance having a conical posterior as sharp as its anterior would be moved after projection just as swiftly as it would be without a sharp conical posterior. But surely the air following could not push a sharp end in this way because the air would be easily divided by the sharpness.
>
> The third experience is this: a ship drawn swiftly in the river even against the flow of the river, after the drawing has ceased, cannot be stopped quickly, but continues to move for a long time. And yet a sailor on deck does not feel any air from behind pushing him. He feels only the air from the front resisting [him]. Again, suppose the said ship were loaded with grain or wood and a man were situated to the rear of the cargo. Then if the air were of such an impetus [that it] could push the ship along so strongly, the man would be pressed very violently between that cargo and the air following it. Experience shows this to be false. Or, at least, if the ship were loaded with grain or straw, the air following and pushing would fold over (*plico*) the stalks which were in the rear. This is all false.[29]

The point of this is to bring to attention that practical activities had affected medieval scholastic thinking. Without doubt Buridan, who was Rector of the University of Paris according to documents in 1328 and again in 1340, took them on board. In the course of his studies of motions of material bodies, Buridan developed the notion of *impetus*. He associated it with the motive force imparted to the body by the agent that set it in motion. Viewed in retrospect, Buridan was coming to grips with the tendency of bodies in motion towards inertia which was to occupy the minds of René Descartes (1596-1650), Gottfried Wilhelm Leibniz (1646-1716) and Newton (1643-1727).[30]

It is noteworthy that Buridan employed *impetus* to discuss whether God is in need of assistance to move celestial bodies. Buridan found that God can do without it – his pervasive sway is sufficient to keep them going.

---

29 J. Buridan, *The Impetus Theory of Projectile Motion*. Transl., intr., and annotated by Marshall Clagett, in Grant (ed.), *Source Book*, pp. 275-76.

30 According to Buridan, 'by the amount the motor moves that moving body more swiftly, by the same amount it will impress in it a stronger impetus'. Ibid., p. 277. This measure of *impetus*, as has been often observed, recalls Newton's *momentum* as defined by the product of mass multiplied by velocity.

Prefiguring Newton, in fact, Buridan proposed that once launched by God such bodies are on their own – possibly for eternity. However, in order to avoid accusations of advancing opinions contrary to the teachings of the Church, Buridan was at pains to stress:

> But this I do not say assertively, but [rather tentatively] so that I might seek from the theological masters what they might teach me in these matters as to how these things take place....[31]

It remains uncertain whether Buridan's *impetus* theory influenced Nicole Oresme when he compared God to a celestial clockmaker allowing the heavens to move, like clockwork on their own.[32]

Oresme is celebrated more often than not for employing geometrical lines and figures to represent and quantify qualities and motions.[33] His novel method is comprehensibly described in Kaye's account as follows:

> [Oresme's] new approach, which he outlined clearly in the first part of his *De configurationibus*, was to construct a dual system of coordinates capable of representing at the same time the intensity of a quality and the extension of the subject in which the quality inhered. In Oresme's scheme, the extension of a given subject in space or time was measured by a base line [longitude], and the intensities of the quality or motion in that subject were represented by perpendicular lines erected on the base line [latitude]. Greater or lesser intensities at various points in the subject were represented by proportionally longer or shorter lines erected on the base line at these points. When drawn, these measuring lines along two coordinates formed two-dimensional surfaces of varying geometrical configurations and sizes.[34]

Thus the configuration of a quality or motion of uniform intensity – the heights of all vertical lines being the same – had to be a rectangle. A quality or motion of uniformly varying intensity was categorised by Oresme as 'uniformly difform', e.g. uniformly accelerated motion. It was represented by a right triangle.

Oresme was interested not only in the geometric representation of quality and motion but also in the measurement of the quantity of the quality of motion, a quantity which he imagined to be equal to the product of intensity and extension. Moreover, he arrived at what amounted to a geometric proof

---

31 Ibid., pp. 277-78.

32 See Crombie, *Augustine to Galileo*, p. 255.

33 See N. Oresme, *The Configurations of Qualities and Motions, including a Geometric Proof of the Mean Speed Theorem*. Transl., intr., and annotated by Marshall Clagett, in Grant (ed.), *Source Book*, pp. 243-52.

34 Kaye, *Economy and Nature*, p. 204.

of the mean speed theorem in which the areas of the rectangle (representing a uniform motion) and the right triangle (representing acceleration) are shown to be equal. The geometric demonstration of the theorem was to influence the analysis of motion for the next 250 years or so.

To view Oresme's two-dimensional procedure as foreshadowing Cartesian analytical geometry is problematic. The simultaneous representation of the extension of a given subject in space or time and the intensity of a quality or motion was not equivalent to the axial system, named the 'Cartesian coordinate system'. Rather, it was a device to demonstrate geometrically that 'quantity (extension) and quality (intension) were bound together in a dynamic proportional relationship'.[35]

What marks Oresme off from medieval scholars concerned with proportionality was, as Grant puts it, his 'fascination with the subject of commensurability and incommensurability in mathematics, physics and cosmology ... [as] evidenced by a number of treatises in which he saw fit to discuss or at least to mention it'.[36]

In fact, Oresme's intellectual interests extended beyond the natural world. Among other works, he wrote a very influential treatise on money and minting known under its abbreviated title *De moneta*.[37] The question suggests itself as to whether Oresme's scientific and economic thinking connect and, if so, in which context. This issue has been perceptively addressed by Kaye in his treatment of areas of historical investigation that are rarely considered together: economic history, the history of economic thought and the history of science on which much of what follows is based.[38]

Monetisation and market developments

Kaye's point of departure is the development of the power and weight of the market place within European feudal society during the twelfth and thirteenth centuries. This growth had to do with the development of towns as centres of trade and handicraft production. Factors in this process, as well as products

---

35 Ibid., p. 206.

36 Grant (ed.), *Source Book*, p. 529.

37 The work's title is given as *Tractatus de origine et natura, iure, et mutationibus monetarum* and is said to have been written sometime between 1355-1360. See Charles Johnson (ed. and transl.), *The "De moneta" of Nicholas Oresme and English Mint Documents* (London: T. Nelson, 1956).

38 Kaye, *Economy and Nature*, pp. 9-10. For a related study, see A. W. Crosby, *The Measure of Reality: Quantification and Western Society, 1250-1600* (Cambridge: Cambridge University Press, 1998).

of it, were widening monetisation and heightening monetary consciousness. These connect to the subject of Oresme's *De moneta*, in which the state's policy of debasement of coinage is critically examined. The devaluation and revaluation of coins were undermining money's role as a socially-accepted general equivalent to the value of commodities. Here was a situation, as Kaye notes, in which society experienced relativity and proportionality on a grand scale. It furnishes the social background to analysis, for instance, of Oresme's and Buridan's relativistic ideas about whether the earth is always at rest in the centre of the universe.

There is a good deal of evidence that the scholastic thinkers of Oxford, Paris and elsewhere were not cloistered intellectuals but engaged in academic and ecclesiastical administration, financial operations and politics. They could not avoid becoming aware of the pervasive social and economic impact of expanding merchant capital on the feudal economy and society.

Indeed, the term 'capital' in its Latin version *capitale* here enters the economic vocabulary for apparently the first time. It appears in a thirteenth-century work entitled *De contractibus usurariis*, composed by Peter John Olivi (Pierre de Jean Olieu), who hailed from the Provence and lived from 1248-1298. He was a member of the Franciscan order and a leader of the faction pressing for a return to the order's original commitment to spiritual things and values. In fact, after Olivi's death the Franciscan superiors deemed his works to be heretical and ordered their destruction.

Though Olivi opposed Franciscan participation in the economic process, he understood that it constituted a core element of social reality, one cardinally affected, as it were, by the exchange of commodities for money. Versed in Roman and canon law-thinking on lending, and familiar with lending practices in the commercial world, Olivi arrived at the notion of *capitale*. At its simplest, it represented the accrued value of borrowed money due to the skilful activities of the investor. Olivi speculated that the relationship between a striker or thrower of an object and the object he sets in motion corresponds to the relationship between an investor and the borrowed money he turns into *capitale*. It has been argued that this analogy amounted to the first formulation of the concept of impetus.[39]

---

39 Here I draw on M. Wolff's stimulating *Geschichte der Impetustheorie* (Frankfurt am Main: Suhrkamp, 1978), pp. 163f.; see also idem, 'Mehrwert und Impetus bei Petrus Johannis Olivi', in J. Miethke and K. Schreiner (eds.), *Sozialer Wandel im Mittelalter* (Sigmaringen: Thorbecke, 1994), pp. 413-23. For a biological Aristotelian approach to impetus theory, see J. Fritsche, 'The Biological Precedents for Medieval Impetus Theory and its Aristotelian Character', *The British Journal for the History of Science*, 44 (2011), 1-27. For

Be that as it may, we have here medieval perceptions of economic and natural phenomena interacting which, on examination, led Kaye to conclude:

> While the physical basis of reality does not change, the social basis of reality does. Over this period society was transformed through the many-faceted social processes of monetization and market development. Every level of society and every layer of institutional growth was affected. Philosophers, from their earliest days as students, were presented with social and economic experiences, rules of conduct, avenues of advancements and models of success unknown by previous generations. As social experiences change, so too do perceptions about how the world functions and is ordered.
> The rigorous intellectual training of the university and its intense atmosphere of challenge and disputation transformed raw perceptions into insights capable of being elaborated through the technical instruments of mathematics and logic. The scholastic habit of synthesis encouraged the linking of insights and principles between different spheres of thought, between the comprehension of the economic order and the comprehension of the natural order. The result was the creation of a new image of nature based on the experience and observation of monetized society: a dynamic, relativistic, geometric, self-ordering, and self-equalizing world of lines. It was upon this model of nature, first imagined in the fourteenth century, that thinkers from Copernicus to Galileo constructed the 'new' science.[40]

Edward Grant's depiction of the medieval societal environment in which knowledge of nature developed appears distinctly narrow in this light. It is insufficient to equate this environment with the translation of Greco-Arabic learning into Latin, the formation of the medieval university and the rise of the theologian/natural philosopher.

Social relations of experimentation

The historic issue is the role of the form of capital, termed merchant capital, in the first phase of the transition from feudalism to capitalism. These are Marxist conceptions advanced for the understanding of the pre-industrial phase of capitalism in Europe. That is, the phase in which the control of capital over the monetary exchange was of greater importance than its

---

further information, see R. W. Hadden, *On the Shoulders of Merchants Exchange and the Mathematical Conception of Nature in Early Modern Europe* (Albany, N.Y.: State University of New York Press, 1994), p. 100.

40 Kaye, *Economy and Nature*, pp. 245-46. See also J. Day, 'Shorter Notices', *The English Historical Review*, Vol. 109/432 (1994), 701.

domination over the production process, constrained, as it was, by urban guilds.

This dynamic certainly emerges from David Abulafia's incisive review of Italian banking in the late Middle Ages. But, along with other historians, Abulafia appears to be reluctant to employ the term 'capitalism' in the context of pre-industrial economies.[41] Be that as it may, it is worth noting that the prominent non-Marxist student of the Scientific Revolution, Steven Shapin, has recourse to the concept of the transition from feudalism to capitalism in Europe between the fifteenth and seventeenth centuries. He identifies this period as one of recognisable societal change, during which mechanical modes of explaining natural phenomena were in the ascendancy. The transformation of the intellectual climate during this period was fuelled by the mounting awareness of mechanical devices in everyday life, symbolised by the clock:

> The allure of the machine, and especially the mechanical clock, as a uniquely intelligible and proper metaphor for explaining natural processes not only broadly follows the contours of daily experience with such devices but also recognizes their potency and legitimacy in ordering human affairs. That is to say, if we want ultimately to understand the appeal of mechanical metaphors in the new scientific practices – and the consequent rejection of the distinction between nature and art – we shall ultimately have to understand the power relations of an early modern European society whose patterns of living, producing, and political ordering were undergoing massive changes as feudalism gave way to early capitalism.[42]

Thus reflecting the course of events, Leibniz acknowledges in 1671 the preference of contemporary natural philosophers for a mechanistic interpretation over the dominant organicist philosophy of Aristotle: 'All modern philosophers desire to explain natural phenomena mechanistically'.[43] Around this time the term 'mechanical philosophy', coined according to all accounts by Robert Boyle, entered scholarly vocabulary.

---

41 D. Abulafia, 'The Impact of Italian Banking in the Late Middle Ages and the Renaissance, 1300-1500', in A. Teichova, G. Kurgan-van Hentenryk and D. Ziegler (eds.), *Banking, Trade and Industry: Europe, America and Asia from the Thirteenth to the Twentieth Century* (Cambridge: Cambridge University Press, 1997), pp. 18, 31.

42 S. Shapin, *The Scientific Revolution* (Chicago, IL and London: University of Chicago Press, 1998), p. 33.

43 'Desiderant omnes philosophi recentiores physica mechanice explicari'. Quoted by H. Mayer, 'Gott und Mechanik Anmerkung zur Geschichte des Naturbegriffs im 17. Jahrhundert', in S. Mattl and others, *Barocke Natur* (Korneuburg: Ueberreuter, 1989), p. 12.

Fig. 6 Portrait of Robert Boyle by Johann Kerseboom (c. 1689).

Among the numerous publications dealing with the work and life of the 'father of the steam-engine', Steven Shapin and Simon Schaffer's *Leviathan and the Air-Pump* has attracted wide attention. They wanted 'to understand the nature and status of experimental practices and their historical products' and they wanted their 'answers to be historical in character'. To that end they write:

> we will deal with the historical circumstances in which experiment as a systematic means of generating natural knowledge arose, in which experimental practices became institutionalized, and in which experimentally produced matters of fact were made into the foundations of what counted as proper scientific knowledge. We start, therefore, with that great paradigm of experimental procedure: Robert Boyle's researches in pneumatics and his employment of the air-pump in that enterprise.[44]

Here the reader is reminded that the two essential and intertwined aspects of science – systematic and quantitative experimentation, on the one hand, and institutionalisation (about which later), on the other – were coming into their own in the seventeenth century.

In fact, the vacuum pump was one of the first of the complex and large machines to be developed for laboratory use (the electrical machine was the

---

44 S. Shapin and S. Schaffer, *Leviathan and the Air-Pump: Hobbes, Boyle, and the Experimental Life* (Princeton, NJ: Princeton University Press, 1985), p. 1. Note: 'The father of the steam engine would be a better title for Robert Boyle than the "father of chemistry"'. A. Rupert Hall, *The Revolution in Science, 1500-1750* (London and New York: Longman, 1983), p. 338.

other).[45] According to Shapin and Schaffer, it was the intellectual production ('matters of fact') by means of a purpose-built scientific machine that made Boyle's experimentation an historical milestone. For one thing, the novel technology for the experimental investigation of air-pressure ('spring of air') clarified why suction pumps would not raise water more than about ten metres – an obstacle for the development of mining in the deep. For another, it gave the lie to the denial of a void in nature – an essential component of Aristotelian physics. Last but not least, from the tabulated measurements of pressures and volumes of air it emerged that their product is the same (1662). Historically, the reciprocal relationship, known as Boyle's Law, is considered to be the first experimental physical law.

There is widespread agreement, however, that the celebrated air-pump experiments were performed by Robert Hooke (1635-1703) and not by Boyle, his employer and backer of many years. All the same, Boyle wished to make a personal point when he stated:

> And though my condition does (God be praised) enable me to make experiments by others' hands; yet I have not been so nice, as to decline dissecting dogs, wolves, fishes and even rats and mice, with my own hands. Nor, when I am in my laboratory, do I scruple with them naked to handle lute and charcoal.[46]

Here Boyle, the wealthy gentleman, clearly signalled that it was not socially demeaning to engage in hands-on experimentation. Moreover, he exhorted scientists

> to disdain, as little as I do, to converse with tradesmen in their work houses and shops ... he deserves not the knowledge of nature, that scorns to converse even with mean persons, that have the opportunity to be conversant with her.

In general, what the natural philosophers endeavoured to comprehend in their workshops (laboratories) was matter in motion. It underlay the processes that craftsmen and artisans were empirically mastering while plying their trades in their workshops. The natural philosophers' growing awareness of the contiguity of these activities weakened the reasoning that underwrote the traditional distinction between nature and art – it was not confined to England.

---

45 S. A. Bedini and D. J. da Solla Price, 'Instrumentation', in Kranzberg and Pursell, Jr. (eds.), *Technology*, Vol. 1, p. 178.

46 This and the following quotation are from J. G. Crowther, *The Social Relations of Science* (New York: The Macmillan Company, 1942), pp. 364-65.

Take the sixteenth-century Czech polymath Tadeáš Hájek z Hájku (1525-1600), also known as Thaddeus Hagecius or Nemicus. He was ennobled on being appointed in 1571 to the post of Chief Medical Officer of Bohemia; moreover, because of his astrological and alchemistic interests he became an influential figure at the court of Emperor Rudolph II (1576-1612) in Prague. In 1585 he published in Frankfurt am Main a small book on brewing: *De cerevisia eiusque conficiendi ratione natura viribus et facultatibus opusculum*.[47]

What lay behind Hagecius's interest in beer? There was the medical and pharmaceutical dimension. It appears that the suggestion to write the booklet came to Hagecius from a personal physician of Rudolph II. But beyond that, there is the growth of a large scale manorial economy in Bohemia to consider. Centred on sheep-farming, brewing and raising fish (carp), as well as on glass and iron making, this economy was directed towards augmenting the declining cash income of the lord of the manor.[48] Thus the lord evolved into a large scale entrepreneur within the feudal system and, unwittingly, into an accessory to its dissolution.

The steps taken by Hagecius to become acquainted with brewing practice recalls Boyle's advice to natural philosophers not to shy away from seeking enlightenment from socially inferior practitioners. Ignorant of brewing, Hagecius consulted with humble brewers who provided him – as he acknowledges – with information that was full, though simple and unsystematic. Poignantly, he regarded the production of beer as a legitimate

---

47 Hagecius has a place in the history of astronomy. First, his findings related to the discovery of a new star in the Cassiopeia constellation (1572). He established that the star must be further from the Earth than the moon and must therefore be a fixed star. This disclosure contributed to the supplanting of the Aristotelian doctrine of two disjointed regions, sublunary and supralunary. He took a similar line in his writing on the comet that appeared in 1577. Moreover, he was behind Rudolph II's invitation to Tycho Brahe (1546-1601) to move to Prague. The life and work of Hagecius is examined by eighteen authors in the collection edited by P. Drábek, *Tadeáš Hájek z Hájku* (Prague: Společnost pro dějiny věd a techniky, 2000). The stock of factual knowledge about Hagecius's life is critically scrutinised by J. Smolka, 'Thaddaeus Hagecius ab Hayck, Aulae Caesarae Majestatis Medicus', in Gertrude Enderle-Burcel, E. Kubů, J. Šoušsa and D. Stiefel (eds.), *"Discourses – Diskurse" Essays for – Beiträge zu Mikuláš Teich & Alice Teichova* (Prague and Vienna: Nová tiskárna Pelhřimov, 2008), pp. 395-412. For a brief introduction in English to the period, including Hagecius's significance, see J. Smolka, 'The Scientific Revolution in Bohemia', in Porter and Teich (eds.), *Scientific Revolution*, pp. 210-39. For an English response to the discovery of a new star in Cassiopeia, see S. Pumfrey, '"Your Astronomers and Ours Differ Exceedingly": The Controversy over the "New Star" of 1572 in the Light of a Newly Discovered Text by Thomas Digges', *The British Journal for the History of Science*, 44 (2011), 29-60.

48 The elements of this economy are echoed in a well-known contemporary German saying: 'Schäfereien, Brauereien und Teich, machen die böhmischen Herren reich'.

field for scientific inquiry and rejected the notion that it was an undignified scholarly pursuit. Clearly, Hagecius did not perceive the worlds of intellectual and manual labour to be separated by an impervious wall.

It is in this context that it is convenient to turn to Shapin and Schaffer's account of Thomas Hobbes's (1588-1679) critique of Boyle's methodology. He took issue with the notion that trustworthy natural knowledge can come by way of (air-pump) experimentation.

For Hobbes, the idea that the Boyleian pneumatic experiments, including crucial measurements of 'the spring of air', could establish the existence of the vacuum was defective in three related ways. To begin with, Hobbes held labouring in a laboratory to be akin to the labours of 'workmen', 'apothecaries', 'gardeners' and, therefore, a class-bound manual activity beneath a philosopher's dignity. Experiment was one thing, philosophy another. By its very nature, knowledge produced by the former was inferior to that generated by the latter.

As to philosophy proper, Hobbes subscribed to plenism, a form of materialism asserting that the world of nature is a plenum made of bodies in motion, there being no room for 'free space', i.e. vacuum.

Finally, Hobbes was in matters of politics a fervent advocate of absolute monarchy. Shapin and Schaffer stress that 'Hobbes's philosophical truth was to be generated and sustained by absolutism'.[49] This serves to remind us that they – unorthodoxly – seek 'to read *Leviathan* as natural philosophy'.[50] Unorthodoxly in the sense that the book has been perceived as a political tract, indeed, as 'the greatest work of political philosophy ever written in English'.[51]

Shapin and Schaffer's exegesis of Hobbes's rejection of vacuism on political grounds is bold:

> ...the argument against vacuum was presented within a political context of use ... He recommended his materialist monism because it would assist in ensuring social order. He condemned dualism and spiritualism because they had in fact been used to subvert order ... For Hobbes the rejection of vacuum was the elimination of a space within which discussion could take place.[52]

The employment of the term 'social order' is not clear. Does it refer to a 'political order' (probably) or to a type of society (feudal, capitalist – less likely)?

---

49 Shapin and Schaffer, *Leviathan*, p. 339.
50 Ibid., p. 92.
51 Cf. promotional description of *Hobbes and Republican Liberty* (Cambridge: Cambridge University Press, 2008), authored by Quentin Skinner.
52 Shapin and Schaffer, *Leviathan*, pp. 99, 109.

At all events, the cited passage and other statements hinge on Shapin and Schaffer's introduction and usage of the term 'space' in a wider sense. Thus 'experimental space' refers to congregations in laboratories for performing/witnessing experiments. Whereas 'philosophical space' or 'intellectual space' bear upon participation in scientific debates and the meetings of groups such as the newly founded Royal Society.

Hobbes denies natural philosophers the right to such activities because they could undermine the sovereignty of absolute monarchy, in his understanding of it. As noted by Shapin and Schaffer:

> Speech of a vacuum was associated with cultural resources that had been illegitimately used to subvert proper authority in the state.[53]

The state in question is absolute monarchy. Absolutism is said to be 'the first international State system in the modern world'.[54] Concurrent to its development, the institutional pursuit of natural knowledge (through scientific societies and journals) was evolving and transcending national boundaries. The contemporaneousness is not accidental: both realms, the political and the scientific, were products of and agents in the transition from feudalism to capitalism. The complexities of the transition over time, involving multiple factors (politics, economy, ideology, wars, etc.), do not allow here for an analysis directed towards a primary cause. But it is worth noticing that during its early phase, systematic and quantitative experimentation became an integral part of the pursuit of natural knowledge opposed by Hobbes, acquiring in the long run a degree of relative autonomy in relation to the state.

Regarding the status of experiment in scientific advancement, Boyle prevailed over Hobbes. What about the status of hypothesis which, famously, was of great concern to Newton? Can it be said that Boyle's empiricism triumphed over Hobbes's rationalism? That would be a simplistic choice with regards to the philosophies of both protagonists. Certainly, it does not reflect the position of the Royal Society, which from its inception (1600) until the early eighteenth century 'was the chief European centre of experimental

---

53 Ibid., p. 91. Here, perhaps, it is appropriate to refer to the attention Quentin Skinner pays to Hobbes's description of liberty in *Leviathan* 'according to the proper signification of the word': 'As soon as we leave the world of nature, however, and enter the artificial world of the commonwealth, we are no longer simply bodies in motion; we are also subjects of sovereign power'. Skinner, *Hobbes*, pp. 162-63. Shapin and Schaffer's book is not listed among Skinner's 'Printed secondary sources'.

54 P. Anderson, *Lineages of the Absolute State* (London: New Left Books, 1974), p. 11.

physics'.[55] As Marie Boas Hall puts it in her close analysis of science in the Royal Society in the seventeenth century:

> However much the Society as a body might hesitate to favour hypothesis, its aim was to establish something more than a collection of random experiment. Mere matter of fact was not valued for itself, but for light it could shed on the Society's object, the establishment of a true philosophy of nature.[56]

55 Rupert Hall, *Revolution*, p. 260.
56 M. Boas Hall, 'Science in the Early Royal Society', in M. Crosland (ed.), *The Emergence of Science in Western Europe* (London and Basingstoke: Macmillan, 1975), pp. 61-2.

3. Institutionalisation of Science

England

The founding of the Royal Society has been linked to the thinking of Francis Bacon (1561-1626) on organised science (about which more in the next chapter). The Royal Society is the premier scientific body in Britain, and some would claim the world, but it is not the oldest.

Two short-lived Italian organisations, Accademia dei Lincei (Academy of the Lynxes) in Rome (1609-1630, or 1603-1651) and Accademia del Cimento (Academy of the Experiment) in Florence (1657-1667), are usually listed as the earliest instances of the modern institutionalisation of science. The distinction of the oldest continuously active scientific society belongs to the Deutsche Akademie der Naturforscher Leopoldina (German Academy of Naturalists Leopoldina). Its founding preceded the incorporation of the Royal Society of London by ten years (1652).[1]

The Royal Society's proximate beginnings go back to 1660, when a group of mathematicians, astronomers and physicians, interested in promoting systematic and experimental knowledge of nature, began meeting weekly at Gresham College in the City of London. The meetings had an informal character but very soon developed into the formal operations of a private

---

1   It was founded as the Academia Naturae Curiosorum on 1 January 1652, by four physicians in the then Imperial Freetown of Schweinfurt (now Bavaria). In 1670 Emperor Leopold I (1640-1705) ratified it as the Sacri Romani Imperii Academia Naturae Curiosorum. In 1687 he sanctioned its prerogative as an independent imperial institution (a rare if not unique case in fragmented Germany), and it became known as the Sacri Romani Imperii Academia Caesareo Leopoldina Naturae Curiosorum. See L. Stern, *Zur Geschichte und wissenschaftlichen Leistung der Deutschen Akademie der Naturforscher "Leopoldina"* (Berlin: Rütten & Loening, 1952). On the neglected Florentine Academy, see M. Beretta, A. Clericuzio and L. M. Principe (eds.), *The Accademia del Cimento and its European Context* (Sagamore Beach, MA: Science History Publications, 2009).

http://dx.doi.org/10.11647/OBP.0054.03

society. It was granted two royal charters by Charles II (1630-1685), in 1662 and 1663 respectively, and was denominated Regalis Societas Londini pro Scientia naturali promovenda. From the start the Royal Society became, and has since remained, a self-governing organisation whose members – Fellows of the Royal Society – are charged a fee for belonging to a social club, as it were.[2]

Fig. 7 View from above of Gresham College, London, as it was in the eighteenth century. By unknown artist, after an illustration in John Ward, *Lives of the Professors of Gresham College* (1740).

The founding of Gresham College (still in existence) goes back to a bequest by Sir Thomas Gresham (?1515-1579). One of the great merchants of the day, he also founded the Royal Exchange of London (1568). Gresham College was primarily an educational institution providing instruction in divinity, astronomy, music, geometry, law, medicine and rhetoric. By 1645, it had also become one of the venues for the coming together of persons interested in discussing scientific problems and in experimenting. In the wake of the Puritan Revolution, some members of the group moved to Oxford (1648-1649) where the Oxford Experimental Philosophical Group at Wadham College

---

2   The origins of the Royal Society have been the subject of a seemingly unending series of studies. Thomas Sprat (later Bishop of Rochester) compiled its first history when it was five years old (1667): *The History of the Royal Society in London*. It has been reprinted and edited with critical apparatus by J. I. Cope and H. W. Jones (St. Louis, MI: Washington University Studies, 1958). It is still of value, as are T. Birch, *The History of the Royal Society of London, for Improving of Natural Knowledge, from its First Rise*, 4 vols. (London: printed for A. Millar, 1756-1757) and C. R. Weld, *A History of the Royal Society*, 2 vols. (London: J. W. Parker, 1848). Charles Webster provides a perceptive picture of the complex origins of the Society in his encyclopaedic *The Great Instauration: Science, Medicine and Reform, 1626-1660* (London: Duckworth, 1975). He discusses the roles of controversial figures such as Robert Boyle, Samuel Hartlib (c. 1600-1652) and Comenius (Jan Amos Komenský (1592-1670)). The 350th anniversary was marked by 22 contributions to B. Bryson (ed. and intr.), *Seeing Further: The Story of the Royal Society* (London: HarperPress, 2010).

was established. According to the mathematician John Wallis (1616-1703), the London group (meeting weekly) agreed that

> (to avoid diversion to other discourses, and for some other reasons) we barred all discourses of divinity, of state-affairs, and of news, other than what concerned our business of Philosophy.[3]

This emphasis on the separation of questions belonging to the province of natural sciences from those concerning theology and politics is significant. The same idea reappears in the widely quoted draft preamble to the Statutes of the Royal Society ascribed to Hooke in 1663:

> The business and design of the Royal Society is – To improve knowledge of naturall things, and all useful Arts, Manufactures, Mechanick practises, Engynes and Inventions by Experiments – (not meddling with Divinity, Metaphysics, Moralls, Politicks, Grammar, Rhetorick, or Logick).[4]

It was after the restoration of the monarchy that Gresham College became the venue for weekly meetings of the Royal Society. Composed at the time, the following four sextains (from a poem containing twenty-four) are of more than passing interest. They document the contemporary awareness of the intertwined worlds of science and overseas commerce. In this context the problem of determining longitude loomed large – not to be solved until 1764 by the Yorkshire carpenter John Harrison. Significantly, what was taught at Oxford and Cambridge is here derided in comparison to the learning of the Greshamites. While the Aristotelian rejection of atomism is laughed at, the Epicurean view that nothing exists besides the atoms and the void is approved of:

> The Merchants on the Exchange doe plott
> To encrease the Kingdom's wealthy trade;
> At Gresham College a learned knott,
> Unparallel'd designs have lay'd,
> To make themselves a corporation,
> And know all things by demonstration.

> This noble learned corporation,
> Not for themselves are thus combin'd,
> But for the publick good o' th' nation,
> And general benefit of mankind.
> These are not men of common mould;
> They covet fame, but condemn gold.

---

3 See Sprat's *History of the Royal Society*, Appendix A.
4 Weld, *Royal Society*, Vol. 1, p. 146.

> This College will the whole world measure,
> Which most impossible conclude,
> And navigation make a pleasure,
> By finding out the longitude:
> Every Tarpaulian shall then with ease
> Saile any ship to the Antipodes
>
> The College Gresham shall hereafter
> Be the whole world's University;
> Oxford and Cambridge are our laughter;
> Their learning is but pedantry;
> These new Collegiates do assure us,
> Aristotle's an ass to Epicurus.[5]

John Wallis was one of the original Greshamites, active in discussing the new scientific advances. Another member of the group was Theodore Haak (1605-1690), a native of the Rhenish Palatinate, who is reputed to have initiated its meetings in 1645, after returning from a diplomatic mission in Denmark in the service of the English Parliament. Also, he was erroneously reported to have invested the group with the name of the 'Invisible College'. As Charles Webster points out, this was a separate body, small and short-lived (?1646-1647), in which Boyle was involved. Webster's scepticism regarding the bearing of the Invisible College on the genesis of the Royal Society stems from Boyle's limitation of 'the principles of our new philosophical college' to 'natural philosophy, the mathematics, and husbandry ... that values no knowledge but as it hath a tendency to use'.[6]

Comenius and the Royal Society[7]

In accordance with this utilitarian thinking, Boyle was drawn to the circle around Samuel Hartlib that was concerned with what may be briefly described as 'practical millenarianism'. He hailed from Prussian Elbing (now Elbląg in Poland), which had strong commercial ties with England. Hartlib's mother was English and, like Haak (who belonged to the same circle), he settled

---

5   Ibid., pp. 79-80, n. 10. Peter Dear points out that Gresham College intended to provide instructions to sailors and merchants in useful arts, and especially in practical mathematical techniques. See his *Revolutionizing the Sciences: European Knowledge and its Ambitions, 1520-1700* (Basingstoke: Palgrave, 2001), p. 53.

6   Webster, *Great Instauration*, p. 61, n. 100.

7   What follows draws in part on my article 'The Two Cultures, Comenius and the Royal Society', *Paedagogica Evropea*, 4 (1968), 147-53.

in England. There he became known as a patron and instigator of projects sustaining human progress through improvements in agriculture, reforms of education and promotion of religious peace. Though sympathetic to Bacon's vision of an organised empirical interrogation of nature, he considered it to be too secular.

It is in this context that Hartlib found the writings of the exiled Czech educational reformer Comenius more congenial. In the wake of the Counter-Reformation, after the defeat of the rebellious estates of Bohemia and their allies in the Battle of White Mountain near Prague (1620), Comenius – a member and priest of the proscribed Unitas Fratrum – in 1628 was forced to emigrate, first to the Polish Leszno. There he composed *Conatuum Comenianorum praeludia etc.*, published at Hartlib's behest in Oxford (1637). The work contained an outline of Comenius's 'pansophic' Christian epistemology, rooted in the amalgamation of the senses, reason and revelation.[8]

Fig. 8 Portrait of an old man thought to be Comenius (c. 1661) by Rembrandt.
Florence, Uffizi Gallery.

---

8 For Descartes's critique, see his letter to Hogelande, August 1638 (?), in A. Kenny (ed. and transl.), *Descartes Philosophical Letters* (Oxford: Clarendon Press, 1970) pp. 59-61. An up-to-date account of Comenius's life and work in English is badly needed. For a brief and convenient treatment in Czech, see J. Polišenský, *Komenský: Muž labyrintů a naděje* [*Komenský: Man of Labyrinths and Hope*] (Prague: Academia, 1996).

Hartlib, who corresponded with Comenius from 1632 and procured some financial aid for him, was the leading spirit behind attempts to move him to England. Arriving on 23 September 1641, it appears that Comenius was to head a body that

> would assimilate the most advanced information in each sphere of knowledge, their collaborative enterprise leading to an encyclopaedic understanding of the material world and the solution of religious controversies among the Protestants, to assist the subsequent reform of the church and education.[9]

The social and political upheavals following the outbreak of the Civil War prevented the realisation of this plan and, in the end, Comenius left England via Holland for Sweden on 21 June 1642. A glimpse of Comenius's vision what this body of scholars could have done may be obtained from the work written during his stay in England. Briefly entitled *Via Lucis*, it was printed in Amsterdam in 1668 and dedicated to the Royal Society. The rational kernel of *Via Lucis* is usually summarised as a plan for a universal language, universal schools and a universal college.[10]

It is no accident that Comenius decided to publish *Via Lucis*, after such a long interval since its composition, within a year of the *History of the Royal Society* being issued by Sprat. While paying respect to Bacon's ideas on collective scientific endeavour, Comenius considered the Royal Society to be *the* body which could and should have carried out his ideas in practice. Yet he clearly perceived that this happened only in part and that there were important differences between the project set out in *Via Lucis* and the route taken by the Royal Society.

Comenius points out that the glorious efforts by the Fellows of the Royal Society, as shown by the published records, have a beautiful affinity with those aims put down in chapter XVI of *Via Lucis*, beginning with paragraph 12. This part of the book deals with Panhistoria. What Comenius had in mind was to compose a critical inductive historical survey of man's knowledge of natural and artificial, moral and spiritual processes in order to sift truth

---

9   C. Webster (ed.), *Samuel Hartlib and the Advancement of Learning* (Cambridge: Cambridge University Press, 1970), p. 29.

10   The full title: *Via lucis, vestigata et vestiganda, hoc est rationabilis disquisitio, quibus modis intellectualis animorum Lux, Sapientia, per omnes omnium hominum mentes, et gentes, jam tandem sub mundi vesperam feliciter spargi posit. Libellus ante annos viginti sex in Anglia scriptus, nunc demum typis exscriptus et in Angliam remissus* (Amsterdam: Apud Christophorum Cunradu, 1668). At the time of writing, I was informed by Cambridge University Library that the copy of the English translation by E. T. Campagnac (Liverpool: Liverpool University Press, 1938) had been missing since 2007. I used the Czech-Latin version instead: *J. A. Comenii Via Lucis J. A. Komenského Cesta světla* (Prague: Státní Pedagogické Nakladatelství, 1961).

from error. [11] In light of this, it is not without interest to learn that some of the Fellows of the Royal Society, according to Sprat, were required

> to examine all Treatises, and Descriptions, of the Natural, and Artificial productions of those Countries in which they would be inform'd … They have compos'd Queries, and Directions, what things are needful to be observ'd in order to the making of a Natural History in general: what are to be taken notice of towards a perfect History of the Air, and Atmosphere, and Weather: what is to be observ'd in the production, growth, advancing, or transforming of Vegetables: what particulars are requisite for collecting a compleat History of the Agriculture, which is us'd in several parts of this Nation. [12]

Sprat then offers particular cases of these inquiries concerning the history of weather, saltpetre, gun-powder and dyeing. [13]

The dedication to the Royal Society in *Via Lucis* is as much laudatory of its pursuits to acquire knowledge in the school of nature (*physics*) as it is exhortatory, asking scientists not to leave out consideration of the study of man with his inborn qualities and faculties (*metaphysics*), and finally of the realm where God is the supreme teacher (*hyperphysics*). Though the senses may apprehend nature, they are useless for the understanding of man which can only be achieved through reason (Comenius calls reason the internal light or the eye of souls). But both senses and reason are equally of no avail when it comes to the comprehension of God's own province, inquiry into which Comenius apparently limits to the 'ultimate' type of question such as: what was it like before the world existed and what will it be like when the world exists no more, and what exists outside this world? According to Comenius the only counsellor and guide within this sphere can be faith in revelation. [14]

The Royal Society's declared policy to exclude all subjects not pertaining to the exploration of nature from the consideration of its members was strictly pursued. But it would be wrong to insist that the problems raised by Comenius had appeared to his contemporaries or those who came after them as pseudoproblems. However, an individual Fellow of the Royal Society had to solve them for himself. The relation of science to religion played an

---

11 Ibid., pp. 242-43.

12 Sprat, *History of the Royal Society*, pp. 155-56.

13 Ibid.: 'A Method For Making a History of the Weather' by Mr. Hook, p. 173; 'The History Of the Making of Salt-Peter' by Mr. Henshaw, p. 260; [Mr. Henshaw], 'The History Of Making Gunpowder', p. 277; 'An Apparatus to the History of the Common Practices of Dying', by Sir William Petty, p. 284. For an account of the Royal Society's early interest in securing natural history, including human society overseas, see J. Gascoigne, 'The Royal Society, Natural History and the Peoples of the "New Worlds", 1660-1800', *The British Journal for the History of Science*, 42 (2009), 539-62.

14 Comenii, *Via Lucis*, pp. 155-56.

important role in Boyle's and Newton's lives, to mention just two of the outstanding figures of the Royal Society. Boyle as a young man came under the direct influence of the Hartlibian/Comenian group, and he retained its belief that revelation and scientific truth are perfectly compatible. The position of Newton, who did not believe in the Trinity, is more complex. Newton thought about theological matters very deeply and had difficulties in reconciling his religious belief with the conception of the mechanical universe.[15]

Comenius was a devout Christian, but it would be misleading to think that his criticisms of the Royal Society derived from the fear that natural sciences would intrude on theology's territory. He merely believed that the answers provided by scientific inquiries amounted only to 'the alphabet of divine wisdom and this was by no means sufficient'.[16] Comenius advanced the view that the natural sciences should set a good example to the politicians and theologians because the principles by which the politicians directed the world were unstable. These principles should be examined and everything untrue should be cast aside. The efforts of the Fellows of the Royal Society to penetrate to the truth on the basis of observations and exact experiments ought to set a wonderful example to those who stood at the helms of human society, either as civil administrators or spiritual guardians of the conscience, encouraging them to fear no examination of their actions.[17] Comenius believed then that statecraft and Christian practice could have learned from the methods employed by natural scientists in their quest for truth. But to leave it at that would not do him justice because he was also concerned with the universality of knowledge and education, and sensed the danger in the Royal Society's one-sided preoccupation with nature.

On the continent many scientific societies and academies, founded after the Royal Society, took up a different attitude and established sections (classes) not only for the study of nature, but also for the humanities (philology, history, philosophy etc.). In this respect, they came much closer to Comenius's project for the organisation of universal knowledge than the Royal Society; at the same time, however, important differences should be noted. Knowledge became specialised and the specialists who met in their respective sections became gradually estranged from their colleagues in other fields – they had nothing or very little to say to each other. These societies

---

15 On this subject chapter IV in J. H. Brooke, *Science and Religion: Some Historical Perspectives* (Cambridge: Cambridge University Press, 1991), is rewarding. See also M. Hunter's intellectual biography *Boyle: Between God and Science* (New Haven, CT and London: Yale University Press, 2009).
16 Comenii, *Via Lucis*, pp. 158-59.
17 Ibid., p. 160.

provided a common roof to a house whose inhabitants largely tended to shut themselves in rooms with no common doors.

The universality which these societies aspired to existed on paper and this became to a great extent the basis of educational theory and practice. This universality was certainly of a different nature to that which Comenius desired. Of course, it could be argued that the societies' method was historically inevitable since scientific knowledge of nature and society could be accumulated only by investigations of relatively isolated facets of natural and social reality. Nevertheless, this was the historical root of the polarisation which eventually formed the basis for the development of the 'two cultures', and which today's educational theory and practice still struggles to come to terms with. In the light of this, Comenius's exhortation to the Royal Society in the preface to *Via Lucis*, and the work itself, merit more than a passing glance.

As for Comenius's visit to England and its disputed relevance to the foundation of the Royal Society, perhaps Rupert Hall's observation is apposite:

> No one has yet succeeded in disentangling completely the personal relations of all those who figure more or less largely in the scientific world of mid-seventeenth-century England. It is likely that most of the forty or fifty men concerned knew something of each other, though mainly associated with one of three chief groups – the Hartlib circle, devoted to social and ethical reform and more occupied with technology than abstract science; or the club of mathematicians, astronomers and physicians meeting at Gresham College; or the Oxford Philosophical Society. There were no barriers between them.[18]

France[19]

The Royal Society was 'Royal' de jure but not de facto. Independent of the state, it was a self-governing organisation of members who were originally asked to pay one shilling for expenses. This was not a hardship as the majority of Fellows were persons of independent means ('gentlemen'). While social rank retained its significance, the Fellows were – as Sprat put it – a *'mix'd Assembly* [Sprat's italics] which has escap'd the prejudices that use to arise

---

18 A. Rupert Hall, *The Revolution in Science*, p. 142.

19 What follows under this and the ensuing heading owes much to my 'Tschirnhaus und der Akademiegedanke', in E. Winter (ed.), *E. W. von Tschirnhaus und die Frühaufklärung in Mittel- und Osteuropa* (Berlin: Akademie Verlag, 1960), pp. 93-107. R. Hahn's *The Anatomy of a Scientific Institution: The Paris Academy of Sciences, 1666-1803* (Berkeley, CA and London: University of California Press, 1971) is still relevant.

from Authority, from inequality of Persons...'[20] The state kept its distance from the pursuits of the London Royal Society, be they practical or theoretical.

This distance did not apply to the Parisian Académie Royale des Sciences, established four years later (1666). Its origins also go back to private gatherings of persons with common interests in expanding the knowledge of nature. It was Jean-Baptiste Colbert's (1619-1683) grasp of the import of scientific investigations which led to King Louis XIV (1643-1715) giving his assent to their institutional embodiment. Judging by his correspondence with a number of scientists, the mercantilist Colbert (surprisingly?) did not adopt a narrow utilitarian attitude towards research. He understood that investigators had to have room for the pursuit of knowledge for its own sake. Therefore, the thirty to forty members of the Académie were freed from financial worries. They received an annual pension and a notable allowance for instruments and other research needs. This enabled the Académie to become a centre of collective and conscious efforts to place scientific pursuits at the service of the state. Considerable attention was paid to the analysis of the modes of handicraft production, including assessment and efficiency of novel mechanical contrivances. The foundation of the Paris Academy and its state-link corresponded to the mercantilist policies vigorously pursued by the French government in the seventeenth century.

Prussia and Saxony

The foundation of the Royal Society in London and the Académie des Sciences in Paris made a strong impression on scientists in other countries, including Germany. Reproaching the members of the Leopoldina for not working creatively, Leibniz noted that the organisation lacked sufficient means and social prestige. Altogether, the social conditions for the development of scientific activities were not propitious in a Germany fragmented in the aftermath of the Thirty Years War.

Indeed it was Leibniz who drew a sombre picture of the plight of the scientific-technical personnel in Germany at the time. While there was no shortage of mechanicians, artisans and experimenters, the governments of the various kingdoms and principalities showed little interest in them. Thus they were really faced with two options: either give up and bury their talents, or leave behind the beggarly living conditions at home and seek

---

20 Sprat, *History of the Royal Society*, pp. 91-2.

opportunities abroad to the detriment of Germany. Leibniz made this point when he submitted a proposal for founding a scientific society in Prussia (approved in 1700 but established in 1711).[21]

Ever intent on promoting the institutionalisation of science in Germany, Leibniz corresponded on this subject (1693-1708) with the mathematician Ehrenfried Walther Tschirnhaus (1651-1708). He also was a constructer of circular and parabolic mirrors with which he succeeded, by focusing sunlight, in obtaining high temperatures.[22]

Fig. 9 Spherical burning mirror by Ehrenfried Walther von Tschirnhaus (1786). Collection of Mathematisch-Physikalischer Salon (Zwinger), Dresden, Germany.

Sharing Leibniz's concerns, Tschirnhaus speculated on the possibility of collective forms of scientific activity in Saxony, particularly after his dream of becoming a pensioner of the Paris Academy came to nothing (1682). Of historical interest is Tschirnhaus's letter to Leibniz (13 January 1693) in which the idea of holding a scientific congress is raised for perhaps the first time:

21 See Leibniz, 'Errichtung einer Societät in Deutschland (2. Entwurf)', in A. Harnack, *Geschichte der königlich preussischen Akademie der Wissenschaften zu Berlin*, Vol. 2 (Urkunden u. Actenstücke) (Berlin: Reichsdruckerei, 1900), p. 23.

22 The extent to which Tschirnhaus's expertise with high temperatures contributed to the discovery of the Meissen porcelain by J. F. Böttger (1682-1719), and thus made him a co-discoverer, has been the subject of much discussion.

Merchants gather at the Leipzig Fair because of their perishable earthly things [*vergänglichen Dinge*]; could not also learned people meet here one day [*einmahl alda*] because of important reasons.[23]

In response to a letter from Leibniz, who was thinking of a self-financing scientific society, Tschirnhaus suggested that the revenues could derive from the exploitation of scientific discoveries, such as his own in optics (27 January 1694). He also specified strict criteria to apply to the selection of the society's members: 1. The aspiring member was truly to employ scientific methods in his work; 2. Science and the desire to obtain the truth was to be his main passion; 3. Self-interest was not to be his main motive; 4. Nor was hankering after personal glory to be his reason for doing research.

At the same time, Tschirnhaus informed Leibniz that he had freed himself from these weaknesses. As proof, he offered to publish his own work anonymously – only under the name of the society. He would not ask for a greater share from the common purse than he was entitled to. He was also prepared to hand over to the common purse all monetary gain he derived from his own optical inventions. But in the end, Tschirnhaus doubted that there were scientists who would match his own example. Leibniz agreed that such persons did not exist. Moreover, he observed that not much was to be expected from 'persons of high rank' (*grosse Herren*), however well-intentioned they were.

History confirmed this view. Tschirnhaus became involved, with the help of Prince Fürstenberg, in setting up a manufactory for producing large mirrors in Dresden (1707). These mirrors did not distort, owing to an innovation of Tschirnhaus in the handling of molten glass. Tschirnhaus hoped that the revenue would secure the foundation of a scientific society. It did not materialise, nor did other plans considered by the two scientists. For example, there was a proposal (not theirs) to establish a central German institution for the improvement of the calendar, out of which an academy was to grow – financed from the proceeds of the monopolised sale of calendars. Tschirnhaus and Leibniz's own exertions to establish an academy in Dresden also came to nothing. While Tschirnhaus remained optimistic, Leibniz tended to succumb to depression, especially in view of obstructions to his plan in Berlin, which came to fruition five years before his death.

---

23 Teich, 'Tschirnhaus', in E. Winter (ed.), p. 101.

Bohemia[24]

Traditionally the rise of organised science in Bohemia is linked to the activities of an informal body known as the Private Learned Society. Founded around 1774, it included the humanities from the outset as an integral part of its concerns, in addition to the natural sciences. This breadth was clearly reflected in the title of its journal, *Abhandlungen einer Privatgesellschaft in Boehmen, zur Aufnahme der vaterlaendischen Geschichte und der Naturgeschichte*,[25] published under the editorship of Ignaz (Inigo) von Born (1742-1791). It appeared six times as an annual between 1775 and 1786. The ideology that informed both the formation of the Society and its journal was Bohemian patriotism to which practitioners of both historical and natural historical disciplines confessed. They were not only a socially mixed assembly (like the Royal Society), but also an ethnically diverse one. The Czech- and German-speaking members considered themselves heirs to a long and honourable tradition of learning effectively inaugurated by the foundation of a *studium generale* in Prague, the first university in Central Europe (1348). Patriotism lay behind their call not only for the exploration of the economic resources but also of the historical past of Bohemia. They agreed that critical analysis and rationalism, so relevant to the scientific study of nature, could be equally successful in the scholarly study of history.

The scientific-technical-economic interests of the founders of the Private Learned Society paralleled those of the founders of the Royal Society and other scientific societies in Europe. These societies were concerned with gaining systematic knowledge of nature for practical use in manufactories and agriculture.

The Oxford English Dictionary locates the first use of 'manufactory' in the year 1618. To all intents and purposes, it meant a place of work in which operations were carried out manually. By the 1690s, economic thinkers (W. Petty, J. Locke) were writing about improving the productivity of labour through its division – machinery was largely limited to watermills and windmills. The classical analysis of the role of the division of labour played

---

24 What follows draws greatly on my previous treatments of the subject. See 'Bohemia: From Darkness into Light', in R. Porter and M. Teich (eds.), *The Enlightenment in National Context* (Cambridge: Cambridge University Press, 1981), pp. 141-63; 247-53; 'Afterword', pp. 215-17.

25 Available at http://babel.hathitrust.org/cgi/pt?id=nyp.33433009935119;view=1up;seq=11

in raising the productivity of labour was offered by Adam Smith (1723-1790) in *The Wealth of Nations* (1776). It influenced Marx to perceive in manufacture the characteristic mode of production of the pre-industrial phase of capitalism. He called it the 'period of manufacture' (*Manufakturperiode*) and thought that it extended roughly from the middle of the sixteenth to the end of the eighteenth century. While this approach informed historians in the former socialist countries, it was largely ignored by Western historiography. Inasmuch as historians variously adopted the concept of 'proto-industrialisation', beginning in the 1970s, they found the Marxist framework of the manufactory stage of industrialization, in certain economically active European states or regions, wanting.[26]

Be that as it may, manufactories needed raw materials, the existence of which could be ascertained through surveys of natural resources. Hence the surveying of natural resources, famously represented in the Swedish context by Carolus Linnaeus (Carl von Linné) (1707-1778), became one of the important tasks the scientific societies set themselves.

The need for an organised scientific survey of the Habsburg dominions had already been proclaimed by Philipp Wilhelm von Hornigk (1638-1712), the leading thinker of Austrian mercantilism, in 1684. Hornigk (Hornick, Hoernigk) recognised the importance of mathematics and mechanics for the development of manufactories. He emphasised that they should use indigenous raw materials. He called for surveys and experiments on the acclimatisation of foreign plants and animals. He also thought it highly desirable to publish a technological encyclopaedia which would explain the significance of physics and mechanics for productive purposes. This task – according to Hornigk – could not be performed by a single person but only by a group of disinterested specialists in various subjects, scientists who would not keep their knowledge to themselves but place it at the public's disposal.[27]

Hornigk's agitation against secretiveness and his request for specialists to combine their scientific and technical knowledge for production and commerce was not accidental. The principle of cooperation based on the

---

26 For an introduction and survey, see S. C. Ogilvie and M. Cerman (eds.), *European Proto-Industrialization: An Introductory Handbook* (Cambridge: Cambridge University Press, 1996); see also M. Berg, *The Age of Manufactures, 1700-1820: Industry, Innovation and Work in Britain*, 2nd ed. (London and New York: Routledge, 1994). For a non-Marxist critique of the theory, see D. C. Coleman, 'Proto-industrialization: A Concept too Many', in his *Myth, History and the Industrial Revolution* (London and Rio Grande, OH: Hambledon Press, 1992), pp. 107-22.

27 [Ph. W. Hornigk], *Oesterreich ueber alles wann es nur will*, 2nd edn ([n.p.]: [n. pub.], 1685), pp. 94f., 99, 261-63.

division of labour, so characteristic of operations in manufactories, was also penetrating the world of science. In some ways artisans and scientists had developed a similar attitude in refraining from divulging what were believed to constitute 'trade' secrets. With the growth of specialised scientific knowledge the need arose for an exchange of observational and experimental results that could be tested and expanded, leading to the foundation of scientific societies and journals. Through them scientific activity became 'socialised' in terms of organisation and also in the sense that its results became public property, available at no cost to those interested in its practical utilisation in industry, agriculture and medicine.[28]

The conditions for Hornigk's suggested association of scientists working for Austria's economic benefit matured only slowly, and it took almost one hundred years before one was founded in Bohemia. The background to the establishment of the Private Learned Society will become clearer in reference to the exploration of the natural resources of Bohemia and the Austrian Salzkammergut, as instigated by Empress Maria Theresa (1740-1780) and her husband Francis of Lotharingia (1745-1765), a leading entrepreneur himself in the 1750s and 1760s. They charged with this task Jan Křtitel Boháč (Johan[n] Tauffer Bohadsch), a professor and leading official of the Prague medical faculty.

One of the distinguished microscopists of his time, Boháč (1724-1768) was also a commercial counsellor to the Bohemian *Gubernium*. Though a university professor, Boháč was not isolated from life and had not the slightest doubt that the development of the natural sciences, the arts and manufacturing formed an inseparable unity. With great clarity he defended the social function of scientific investigations against those who tended to underrate it.

In the eighteenth century, under the influence of the much-travelled Linnaeus, systematics came to occupy a central place in natural history. Sometimes these endeavours degenerated into aimless classifications of plants, animals and minerals for their own sake. Boháč condemned such tendencies, holding that the classification of natural objects should be a means towards utilising them in material production. His comprehensive approach led him to appreciate the dependence of manufacture on agriculture. For instance, his concern with the cultivation of woad for animal feeding and for dyeing indicated the connection between scientific, technical, economic and political aspects of his work.[29] It was to be crowned by a comprehensive survey of

---

28 See K. Marx, *Capital* (London: George Allen & Unwin, 1938), Vol. 1, p. 383.

29 J. T. Bohadsch, *Beschreibung einigen in der Haushaltung und Faerbekunst nutzbaren*

the plant, animal and mineral wealth of Bohemia. However, because of Boháč's untimely death, it remained as a manuscript that has since been lost. There can be little doubt about the social and economic impetus that turned Boháč and others to apply their expert knowledge of the properties and processes of nature, inanimate and animate, to practical fields, including the systematic survey of the natural wealth of Bohemia. But an individual, unaided financially, could hardly complete this work alone.

A body of individuals could, and it might have been expected that Prague University would eventually house such a body because it counted among its members prominent scientists interested in the practical use of natural knowledge, like Boháč or the able mathematician and physicist Joseph Stepling (1716-1778). At the order of the Empress Maria Theresa, a kind of university scientific society presided over by Stepling had been established in 1753. University teachers used to meet and hold lectures, but within less than a decade the society ceased to function, possibly due to the antagonism of the Jesuit order still in control of university life and imbued by thinking derived from the Aristotelian-Thomist synthesis.[30]

To complement this account, mention should be made of the earliest scientific society in the Czech Lands and, indeed, in the Habsburg Empire. It was the short-lived *Societas eruditorum incognitorum in terris austriacis* at Olomouc (the former capital of Moravia). It was founded in December 1746, with the backing of Maria Theresa, by Joseph von Petrasch (1714-1772), a former *aide-de-camp* to Eugene of Savoy. The Society of Unknown Scholars arose from informal gatherings of laymen, clergymen and military officers interested in discussing literary and scientific developments at home and abroad in an atmosphere free of the limitations imposed by the Jesuits on the

---

Kraeutern, die er in seinen durch drey Jahre unternommenen Reisen im Königreich Böheim entdecket hat (Prague: Franz Ignatz Kirchner, 1755); *Abhandlung vom Gebrauch des Waides in der Haushaltung* (Prague: [n. pub.], [n.d.]); *Dienst- und Nutzbarer Patriotischer Vorschlag, wienach dem Königreich Böheim ein ungemeiner Vortheil von sonderbarer Beträchtlichkeit jährlich zuwachsen könnte* (Prague: [n. pub.], 1758). See Z. Frankenberger, 'Jan Křtitel Boháč: Jan Křtitel Boháč: život a dílo', *Věstník Královské české společnosti nauk*, 12 (1950), 1-122.

30 Except for scattered remarks in eighteenth-century records, there is little solid information on these meetings, variously called *consessus philosophicus, consessus philosophici* and *consessus literarii*. As to Jesuits in Bohemia, it is necessary to differentiate between the unprogressive attitude of the order delaying the advance of science, and the progressive role of its individual members in furthering and participating in astronomical, mathematical and physical inquiries (e.g. J. Stepling). See also E. Winter, 'Die katholischen Orden und die Wissenschaftspolitik im 18. Jahrhundert', in E. Amburger, M. C. Cieśla and L. Sziklay (eds.), *Wissenschaftspolitik in Mittel- und Osteuropa* (Berlin: Camen, 1976), pp. 85-96.

spiritual life of the fortress and university town. Containing only reviews and no original contributions, two volumes of its journal *Monathliche Auszuege Alt, und neuer Gelehrten Sachen* were issued at Olomouc in 1747 and 1748. Apparently to avoid censorship, the third volume of the journal was printed outside Austria (1750) and afterwards its publication ceased.[31]

About two decades elapsed before the idea of a scientific society was taken up again by the well-known mineralogist Ignaz von Born. Writing to his friend Count F. J. Kinsky(ý) (1739-1805), Born emphasised that nobody had thought of setting up a learned society for the exploration of Austria's widespread territory, to assemble the observations made by naturalists and scientists. This is evidence that a scientific society, whether centred on Austria as a whole or restricted to Bohemia, did not exist before 1774.

Born criticised the aristocracy for its lack of comprehension of the utility of natural history. He stressed that those who took interest in it and had the ability to work creatively did not possess the means to explore nature. He explicitly mentioned the case of Boháč, who on his travels had collected natural objects at his own expense and on his death left his wife penniless. Whereas, according to Born, the nobility had the means but did not encourage people of talent to investigate the natural wealth of the monarchy. Furthermore, in his letter he elucidates the usefulness of science to the economy, the state, the church, the doctor and the poet.[32]

He was particularly concerned with the perniciousness of not making scientific observations and technical discoveries available to all, under the cloak of state secrecy. Born here was condemning an official practice which had already almost landed him with the charge of treason. In 1771 he had published N. Poda's (1723-1798) descriptions of machines used in the mining district of Banská Štiavnica, one of the classical texts on eighteenth-century mining in Central Europe.[33] At that time, he occupied the post of assessor of

---

31 There is as yet no reliable treatment of the subject. Under these circumstances what can be said, in general, is that the *Societas incognitorum* embodied an effort to organise scientific and cultural life at an early stage of the Enlightenment in the Habsburg dominions, but the social, intellectual, local and personal circumstances that engendered its birth were not adequate to keep it alive. See E. Wondrák, 'Die Olmützer "Societas incognitorum". Zum 225. Jubiläum ihrer Gründung und zum 200. Todestag ihres Gründers', in E. Lesky, D. S. K. Kostić, J. Matl and G. v. Rauch (eds.), *Die Aufklärung in Ost- und Südosteuropa* (Cologne and Vienna: Böhlau, 1972), pp. 215-28.

32 *Schreiben des Herrn Ignatz von Born … an Herrn Franz Grafen von Kinsky, Ueber einen ausgebrannten Vulkan bey der Stadt Eger in Boehmen* (Prague: Gerle, 1773), pp. 1-3, 11-16.

33 N. von Poda, *Kurzgefasste Beschreibung der, bey dem Bergbau zu Schemnitz in Nieder-Hungarn, errichteten Maschinen etc.* (Prague: Walther, 1771).

the Mint and Mining Head Office in Prague, from which he chose to resign. The defence of open scientific communication was crucial to Born's drive to organise scientific life in Bohemia between 1770 and 1776, and afterwards in Vienna and, indeed, on an international scale.[34]

Major-General Franz Joseph (František Josef) Kinsky was descended from one of the great Czech aristocratic houses. A keen geologist and educationalist, he eventually became the head of the Military Academy at Wiener Neustadt. He supported Born's vision of putting scientific life in Bohemia on an organised basis for economic, technical and educational reasons. Together with Born and aided by the head of the *Gubernium*, Prince Karl Egon Fürstenberg, he was instrumental in founding the Natural History Museum (1775) and bringing into being the Prague University Library (1777), of which he became the first director.[35]

Kinsky shared Born's concern that the aristocracy as a social class was apt to regard science and technology with disdain. In a letter to Born published in the first volume of *Abhandlungen* (1775), Kinsky complained that the nobility were not properly informed that the administration of their domains required knowledge of natural and agricultural sciences. In his answer Born wrote that a mineralogical and geographical description of Bohemia was needed, adding that there were only a few mineralogists available. According to Born, they ought to follow the example of Saxony, where specialists financed by public funds were preparing a mineralogical map.[36]

The Private Learned Society's transformation into a public institution occurred when it became the Bohemian Society of Sciences in 1784 and the Royal Bohemian Society in 1790. The problems which the scientists in Bohemia tried to solve, especially those associated with the Royal Bohemian Society of Sciences or within its orbit during the first period of its existence, were closely related to the idea of a scientific survey of Bohemian natural resources.

The Society approached the problem of a scientific survey of Bohemia basically from two angles. It launched prize essay competitions and organised

---

34 The definitive study of this leading figure of the Enlightenment in the Habsburg monarchy still remains to be written. See H. Reinalter (ed.), *Die Aufklärung in Österreich. Ignaz von Born und seine Zeit* (Frankfurt am Main; Bern; New York; Paris: Lang, 1991).

35 Kinsky has received little serious attention. For an appreciation, see J. Haubelt, 'František Josef Kinský', *Věstník Československé akademie věd*, 78 (1969), 560-77.

36 'Schreiben des Herrn Grafen von K... an Herrn von Born ueber einige mineralogische und lithologische Merkwuerdigkeiten', *Abhandlungen*, 1 (1775), 243-52; 'Antwort des Herrn von Born, auf das Schreiben des Herrn Grafen von K.....', ibid., 1 (1775), 253-63.

expeditions for the purpose of surveying various regions of Bohemia. The aim of these endeavours was to collect a large amount of scientifically verified information for a map of Bohemia.

The members of the Society embarked upon this plan because they were convinced that the development of manufacturing depended above all on knowledge of domestic economic resources. However, the social, financial and personal situation did not favour the transformation of this awareness into reality. For one thing, the continuing feudal relations and undeveloped capitalist relations effected negative progress in agriculture and industry. For another, the Bohemian Society of Sciences was in continuous financial difficulties which were not alleviated despite the support of a few interested aristocrats. In addition, the number of individuals able to perform a large-scale survey of the country was then small. The Society included amongst its members (nearly all non-nobles) the most distinguished scholars in Bohemia, but that amounted to no more than a few persons. As a consequence it succeeded only partially in achieving its aim.

4. Truth(s)

Collective truth: Bacon

That the collaborative interrogation of the natural world, promoted by the Royal Society in its early days, could be more productive than individual endeavour was brought home to the learned world by Francis Bacon. His depiction of the fictional Solomon's House in the *New Atlantis* (1627) was to serve as a prototype of organised scientific activities for the satisfaction of human needs.

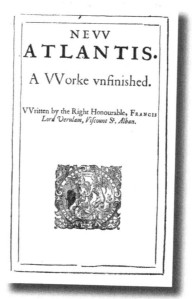

Fig. 10 Title page of *New Atlantis* in the second edition of Francis Bacon's *Sylva sylvarum: or A naturall historie. In ten centuries* (London: William Lee at the Turks, 1628).

http://dx.doi.org/10.11647/OBP.0054.04

This imaginary institution was made up of 36 investigators engaged in collecting information and producing knowledge of nature, including natural histories and surveys of natural resources. Fundamentally, they were collectively concerned with understanding how matter in motion works, and with using it for human progress:

> The End of our Foundation is the knowledge of Causes, and secret motions of things; and the enlarging of the bounds of Human Empire, to the effecting of all things possible.[1]

The division of labour between the members of Solomon's House was functional: they were assigned to nine research groups, charged to perform specific jobs. The largest one comprised twelve travellers who visited foreign countries. Their name, the 'Merchants of Light', is double-edged since they were engaged in clandestine scientific intelligence. As for the eight "home-based" groups, they were made up of three persons each, and named no less arrestingly.

The 'Depredators' gleaned information about experiments from books. The 'Mystery-men' were concerned with tracking down trade secrets (techniques). The 'Pioneers or Miners' explored new avenues by designing novel experiments. The 'Compilers' summarised the results of the former four groups 'for the drawing of observations and axioms out of them'. The 'Dowry-men or Benefactors' were occupied with material benefits stemming from the 'experiments of their fellows'. The 'Lamps' – after meetings and consultations with the 'whole number' and adjudging what the collective managed to achieve – proposed 'new experiments, of a higher light, more penetrating into nature than the former'. The 'Inoculators' performed them and reported on them. Finally, the 'Interpreters of Nature' connected the findings and expounded them in general terms, including axioms and aphorisms.

There were twenty or so laboratories in Solomon's House serviced by a great number of male and female servants and attendants. Additionally, there were novices and apprentices in order to secure continuity of research. As Krohn points out, the description of the laboratories in Solomon's House, including the equipment and work, must have filled a seventeenth-century reader with awe. While there is the danger of ascribing anachronistic foreknowledge, a

---

1   This and further citations are from Bacon's description of Solomon's House, reprinted as an Appendix in B. Farrington, *Francis Bacon: Philosopher of Industrial Science* (London: Macmillan; New York: Haskell House, 1973), pp. 179-91. A classical study which, along with P. Rossi, *Francis Bacon: From Magic to Science* (Chicago, IL: University of Chicago Press, 1968), influenced W. Krohn's insightful (German) *Francis Bacon* (Munich: C. H. Beck, 1987).

twenty-first-century reader should have no difficulty in discerning what amounts to experimentation in hybrid sterility when he reads:

> We find means to make commixtures and copulations of different kinds; which have produced many new kinds, and them not barren, as the general opinion is.

Historically, the celebrated instance is the mule – the sterile offspring of a male donkey and a female horse.

Briefly, Bacon's notion of scientific procedure embraced collaborative gathering of empirical/experimental facts combined with inductive generalisations – all in the service of man. This was a radical departure from the attitude going back to Plato and Aristotle, who held 'that the pursuit of knowledge was an end in itself'.[2] In this context, William Harvey's reputed, and frequently quoted, gibe that Bacon wrote philosophy like a Lord Chancellor is beside the mark.

Nevertheless there is a historiographical issue regarding the interaction of law, politics and epistemology in Bacon's approach to understanding nature. Demonstrably, Bacon's legal and statesmanly background had something to do with his analogy between 'axioms' and 'laws'. But it would be a step too far to trace Bacon's position solely to this source. In fact, there are good reasons to connect Bacon's 'laws of nature' with the empirical 'rules' guiding practice in dominant sectors of the economy in seventeenth-century Britain: agriculture and urban and countryside handicrafts. There are grounds, *pace* Farrington, to regard Bacon as a philosopher of pre-industrial science.[3]

On the face of it, Solomon's House was a self-governing and self-financing institution for co-operative research. The funds to carry out the extended research programme essentially derived from home-bred manufacturing and commercial activities and inventions. Presumably financial self-sufficiency enabled the scientists to maintain distance in respect to the state. This is revealed in a remarkable passage towards the end of the description of Solomon's House. The decision to publish results or to keep them under wraps, even *vis-à-vis* the state, was collegiate:

> And this we do also: we have consultations, which of the inventions and experiences which we have discovered shall be published, and which not:

---

2 G. E. R. Lloyd, *Early Greek Science: Thales to Aristotle* (London: Chatto & Windus, 1970), p. 132.

3 W. Krohn, 'Social Change and Epistemic Thought (Reflections on the Origin of the Experimental Method)', in I. Hronszky, M. Fehér and B. Dajka, *Scientific Knowledge Socialized* (Boston Studies in the Philosophy of Science, Vol. 108) (Dordrecht, Boston and London: Kluwer, 1988), pp. 165-78.

and take all an oath of secrecy, for the concealing of those which we think fit to keep secret: though some of those we do reveal sometimes to the state, and some not.

It is said that Hobbes served Bacon as a secretary for a while. Whether the claim is true or false, Hobbes certainly had a different understanding of freedom of action for scientists than his (alleged) master and mentor.[4]

Personal truth: Descartes

Descartes is traditionally presented as Bacon's antipode with respect to epistemology. One way to demarcate the two thinkers (following Koyré) is to say that whereas Bacon was interested in the 'order of things', Descartes searched for the 'truth of ideas'. This distinction is supported by autobiographical narratives (in as much they are credible) from which both men emerge as persons confident of their own qualifications for procuring truthful knowledge.

There is one autobiographical piece by Bacon, as noted by Farrington, composed as a Preface (1603) to *On the Interpretation of Nature*, which was never written in its projected form. This is how Bacon details his inborn attributes to be a seeker of truth regarding the 'order of things':

> For myself, I found that I was fitted for nothing so well as for the study of Truth; as having a mind nimble and versatile enough to catch the resemblances of things (which is the chief point), and at the same time steady enough to fix and distinguish their subtler differences; as being gifted by nature with desire to seek, patience to doubt, fondness to meditate, slowness to assert, readiness to reconsider, carefulness to dispose and set in order; and as being a man that neither affects what is new or admires what is old, and that hates every kind of imposture. So I thought my nature had a kind of familiarity and relationship with capital Truth.[5]

At the heart of Bacon's approach, as previously noted, was the pursuit of knowledge for human welfare. Autobiographically, it is described as follows:

> But above all, if a man could succeed, not in striking out some particular invention, however useful, but in kindling a light in nature – a light which should in its very rising, touch and illuminate all the border-regions that confine upon the circle of our present knowledge; and so spreading further and

---

4   H.-D. Metzger, *Thomas Hobbes und die Englische Revolution 1640-1660* (Stuttgart-Bad Constatt: Frommann-Holzboog, 1991), p. 298.

5   Farrington, *Francis Bacon*, pp. 54-5.

further should presently disclose and bring into sight all that is most hidden and secret in the world, – that man (I thought) would be the benefactor indeed of the human race, – the propagator of man's empire over the universe, the champion of liberty, the conqueror and subduer of necessities.[6]

As to Descartes, his best known work *Discourse on the Method* (1637) offers an autobiographical account of the evolution of his thinking. It supports Koyré's broad demarcation between the two thinkers. Not least because it contains the celebrated deduction whereby Descartes established (to his satisfaction) his own existence: 'I am thinking (*je pense*), therefore I exist'.[7]

Trust in personal truth and distrust of collective truth motivated Descartes to reconstruct human knowledge:

> But from college days I had learned that one can imagine nothing so strange and incredible but has been said by some philosopher; … at the same time, a majority of votes is worthless as a proof, in regard to truths that are even a little difficult of discovery; for it is much more likely that one man should have hit upon them for himself than that a whole nation should. Accordingly I could chose nobody whose opinion I thought preferable to other men's; and I was as it were forced to become my own guide.[8]

As noted, Bacon's path to generalisations was through inductive reasoning sustained by ensuing empirical/experimental evidence. Descartes argued for the opposite procedure based on deductive reasoning from a first principle which he traced to God, the supreme legislator. It was not that he was not aware of the part played by 'practical philosophy' in the replacement of philosophical speculations taught by Schoolmen:

> For I thus saw that one may reach conclusions of great usefulness in life, and discover a practical philosophy in the place of the speculative philosophy taught by the Schoolmen; one which would show us the energy and action of fire, air, and stars, the heavens, and all other bodies in our environment, as distinctly as we know the various crafts of our artisans, and could apply them in the same way to all appropriate uses and thus make ourselves masters and owners of nature....[9]

---

6 Ibid., p. 54.
7 *Discourse on the Method, etc.*, in E. Anscombe and P. T. Geach (ed. and transl.), *Descartes: Philosophical Writings*, with an Introduction by A. Koyré (Sunbury-on-Thames: Nelson, 1976), p. 31. It has been noted that in *Meditations on First Philosophy* (1641), Descartes dispenses with 'therefore'. See ibid., pp. 67f.
8 Descartes, *Discourse*, pp. 18-9.
9 Ibid., p. 46.

It was not that he regarded experiments and observations as unnecessary (he dissected animals), but that they

> often deceive us, so long as the causes of the more common are still unknown; and the conditions on which they depend are almost always so special and so minute that it is very hard to discern them. My general order of procedure on the other hand has been this. First, I have tried to discover in general the principles or first causes of all that exists or could exist in the world. To this end I consider only God, who created them, and I derive them merely from certain root-truths that occur naturally to our minds. Then I considered the first and most ordinary effects deducible from these causes... and then I tried to descend to more special cases.[10]

These special cases included anatomical investigations conducted from the early 1630s to the late 1640s. While reinforcing Descartes's deciphering of living processes along micro-mechanical lines, they raised the question of the relation of the material machine-like body to the incorporeal immortal soul.

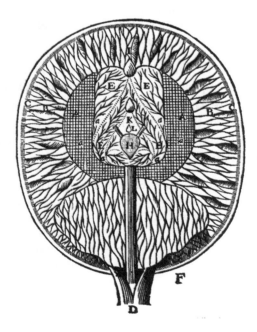

Fig. 11 A diagrammatic section of the human brain
by René Descartes, in his *Treatise of Man* (1664).

---

10 Ibid., p. 47. Noteworthy is that while Marx in *Capital* refers to Descartes's counterpoint of practical and speculative philosophy, he also characterises Bacon and Descartes as philosophers of the pre-industrial phase of capitalism ('period of manufacture'). See K. Marx, *Capital* (London: George Allen & Unwin, 1938), Vol. 1, p. 387, n. 2.

The received wisdom is that Descartes, demarcating sharply between them, parented mind-body dualism. True, in addressing the issue of how mind and body interacted Descartes famously visualised the pineal gland as the seat of the rational soul. There it was supposed to receive external physical stimuli and to direct movements. Descartes's choice of the pineal gland was guided by the notion (erroneous as it turned out) that it is an unduplicated anatomical structure present only in human brains. But this did not imply or signify that Descartes was doubting or even giving up the belief in the existence of God and the immortality of the soul.

Mind and body

Shortly before his death, Descartes reiterated in the Preface to the unfinished *Description of the Human Body* (1647-1648) the importance of personal cognition: 'There is no more fruitful occupation than to try to know itself'. While granting the practical (medical) value of knowing 'the nature of our body', he was 'not attributing to the soul functions which depend only on the body and on the disposition of organs':

> When we make the attempt to understand our nature more distinctly, however, we can see that our soul, in so far as it is a substance distinct from body, is known to us solely from the fact that it thinks, that is to say, understands, wills, imagines, remembers, and senses, because all these functions are kinds of thoughts. Also, since the other functions that are attributed to it, such as the movement of the heart and the arteries, the digestion of food in the stomach and such like, which contain in themselves no thought, are only corporeal movements, and since it is more common for one body to be moved by another body than by the soul, we have less reason to attribute them to the soul than to the body.[11]

To Descartes the cognitive functions of the immaterial soul remained of central concern. Behind them lurked the problem of mind-body dualism, penetratingly raised by Elizabeth, Princess of Bohemia, in her correspondence with Descartes (1643-1646). That is, how an unextended, immaterial soul can move material things. This, and other issues relating to emotions (passions) taken up in the correspondence, undoubtedly prompted Descartes to apply himself deeply to the dichotomy of mind-body dualism.[12]

---

11 Descartes, 'The Description of the Human Body, etc.', in S. Gaukroger (ed.), *Descartes: The World and Other Writings* (Cambridge: Cambridge University Press, 1998), pp. 170-71.
12 Elizabeth, Princess of Bohemia (1618-1680) was the daughter of Frederick V, the Elector of the Rhenish Palatinate and Elizabeth Stuart, daughter of James I and VI, King of

The dichotomy was not resolved by a convoluted 'doctrine of substantial union of mind and body'.[13] It was premised on the existence of two forms of mind. The 'embodied mind' was tied to corporeal organs, perceiving, remembering, imagining, etc. As for the 'disembodied mind', Descartes needed it on account of his belief in the personal immortality of the soul after death. Descartes wanted to have his cake and eat it, that is, to have both a mind dependent on matter and a mind independent of matter. As put by Engels, the basic philosophical issue is not *passé*:

> The great basic question of all philosophy, especially of more recent philosophy, is that concerning the relation of thinking and being... The answers which the philosophers gave to this question split them into two great camps. Those who asserted the primacy of spirit to nature and, therefore, in the last instance, assumed world creation in some form or other ... comprised the camp of idealism. The others, who regarded nature as primary, belong to the various schools of materialism.
>
> These two expressions, idealism and materialism, originally signify nothing else but this....[14]

England and Scotland. Frederick V accepted the Crown of Bohemia but was driven out of the country in the wake of the failed uprising/rebellion of the Bohemian estates (1618-1620) that effectively set in motion the Thirty Years War. In the history textbooks he is referred to as the 'Winter King' (4 November 1619 to 8 November 1620). For a selection of letters between Princess Elizabeth and Descartes, see E. Anscombe and P. T. Geach (ed. and transl.), *Descartes: Philosophical Writings*, with an Introduction by A. Koyré (Sunbury-on-Thames: Nelson, 1976), pp. 274-86. The Elizabeth-Descartes correspondence and relationship are discussed in S. Gaukroger's invaluable *Descartes: An Intellectual Biography* (Oxford: Oxford University Press, 1995), pp. 384f.

13 Gaukroger, ibid., pp. 388f.

14 F. Engels, 'Ludwig Feuerbach and the End of Classical German Philosophy', in K. Marx and F. Engels, *Selected Works in Three Volumes*, Vol. 3 (Moscow: Progress Publishers, 1973), pp. 345-46. For a contemporary and stimulating treatment of the topic, see K. Bayertz, 'Was ist moderner Materialismus', in K. Bayertz, Myriam Gerhard and W. Jaeschke (eds.), *Weltanschauung, Philosophie und Naturwissenschaft im 19. Jahrhundert*, Vol. 1: *Der Materialismus-Streit* (Hamburg: Meiner, 2007), pp. 50-70.

5. The Scientific Revolution: The Big Picture

Was science born/invented and, if so, when?

'There is no such thing as the Scientific Revolution and this is a book about it'.[1] With this somewhat baffling sentence Steven Shapin begins his scrutiny of the movement which, as I argue, came into its own in certain European countries in the sixteenth and seventeenth centuries. That is, a universal mode of producing natural knowledge materialised, one that did not exist anywhere before and that is still practised now. It merits to be designated as *the* Scientific Revolution.

My approach to this thing called the 'Scientific Revolution' is not part of the historiographical mainstream. Indeed since the 1980s, a view has been gaining ground to the effect that the concept itself should be blotted out of the historiographical landscape altogether. This is certainly the message conveyed by Andrew Cunningham and Perry Williams in a thoroughly revisionist paper that works from the unchallengeable principle, advanced in the first sentence, that 'Like it or not, a big picture of history of science is something which we cannot avoid'.[2] The point deserves to be taken up.

The concept of the so-called 'scientific revolution' [*sic*], the authors contend, proved double-edged. It was useful, before and after the Second World War, insofar as it became a unifying concept in the history of science, and helped

---

1   S. Shapin, *The Scientific Revolution* (Chicago, IL and London: University of Chicago Press, 1998), p. 1.
2   A. Cunningham and P. Williams, 'De-centring the 'Big Picture': The Origins of Modern Science and The Modern Origins of Science', *The British Journal for the History of Science*, 26/4 (1993), 407. For the citations that follow, see 409-10, 412, 425, 427-28.

http://dx.doi.org/10.11647/OBP.0054.05

to establish the subject as an independent scholarly and teaching discipline. But it became obsolete insofar as it promoted a big picture that anchored the origins of modern science in the seventeenth century – as distinct from medieval and ancient science. While conceding that 'there is no reason why science should not have originated at that time', they believe, 'this happens not to be the case'. Instead, they suggest 'the period 1760-1848 is a much more convincing place to locate the invention of science'.

In preferring the period 1760-1848, they acknowledge that they were influenced by E. J. Hobsbawm's *The Age of Revolution: 1789-1848* (1962). This book is about the transformation of the world impelled by the French Revolution of 1789 and the contemporaneous (British) Industrial Revolution. Cunningham and Williams enlarge the narrative by adding a German dimension which they call the 'post-Kantian intellectual revolution'. As a result of these three simultaneous and linked revolutions, they argue, 'a new middle-class was constituted wielding the political power, the industrial power and the intellectual power'. It is to these social transformations, as one aspect of the Age of Revolutions, that Cunningham and Williams trace the origins of science – 'with England as partial exception, in that changes there took place later and more gradually than in Continental Europe'. To sum up, they state:

> historical relationships over the last twenty years enable us to identify the Age of Revolutions as the period which saw the origin of pretty well every feature which is regarded as essential and definitional of the enterprise of science: its name, its aim (secular as distinct from godly knowledge of the natural world), its values (the 'liberal' values of free enquiry, meritocratic expert government and material progress), and its history.

The summary of Cunningham and Williams's thesis raises questions that need to be pursued further. All the more so since they justify 'the invention of science' in the period 1760-1848 as follows:

> This term 'invention', which is our preferred term, helps to fix the revised view of science as a contingent, time-specific and culture-specific activity as only one amongst the many ways-of-knowing which have existed, currently exist, or might exist; and for this reason the phrase which we propose for the fundamental changes which took place in this period is 'the invention of science'. And we can of course now drop the qualifier 'modern', since the term 'science' can only be properly applied in our own time, the modern era. What we are speaking of is therefore not the origins of modern science, but the modern origins of science.

'Natural philosophy' and 'science': what's in a name?

The history of science is nothing if it is not about how our knowledge of nature (variously termed) came to be what it is today.[3] In the English context the substitution of 'science' for 'natural philosophy', made around 1800, should not obscure the fact that the terms cover common pursuits.[4]

Take the anti-Trinitarian Newton, deeply concerned with God, and his great successor P.-S. Laplace, who (reputedly) dismissed Him peremptorily. Whatever the grounds and impulses for their quests, godly or ungodly, both inquirers were engaged in the same enterprise: the comprehension and explanation of natural phenomena. The implication of Cunningham and Williams's proposition is that Newton's work was neither 'modern' nor 'scientific'. Patently, this is problematic, since Newton's theory, in John North's telling phrase, 'even on a cosmological scale ... is not yet worn out'.[5]

To distinguish between 'natural philosophy' and 'science' along 'godly' and 'secular' lines is to adopt a narrow standpoint. In 1931, the Soviet physicist Boris M. Hessen (1893-1936), inspired by the Marxist approach to history, first suggested that to understand Newton's work and worldview required one to see them as the product of a specific period. He characterised this historical moment as the period of the feudal economy's disintegration, of the development of merchant capital, of international maritime relationships and of heavy (mining) industry. Hessen presented his ideas in a paper, given at the 2nd International Congress of the History of Science and Technology

---

3 I draw on Eric Hobsbawm's thesis 'that history is engaged on a coherent intellectual project, and has made progress in understanding how the world came to be the way it is'. See E. Hobsbawm, *On History* (London: Weidenfeld & Nicolson, 1997), p. x.

4 According to the *Oxford English Dictionary*, the use of 'science' in the English language goes back to the fifteenth century. By 1830, when Charles Babbage (1792-1871) publishes his highly critical *Reflections on the Decline of Science and on Some of its Causes*, the word had come to mean what it means today. According to the same source, the 'cultivator of science' became a 'scientist' thanks to W. Whewell, in 1840. Babbage played a major role in the foundation of the British Association of Science (1830). It took its cue from the annual Versammlung Deutscher Naturforscher und Ärzte (Assembly of German Naturalists and Physicians). Note that the literal translation of 'Naturforscher' is more like 'investigator' or 'explorer' of nature. 'Naturwissenschaftler' as an equivalent to the English 'scientist' is of later date. For a nuanced 'Note about "Science"', see the informative D. E. Harkness, *The Jewel House: Elizabethan London and the Scientific Revolution* (New Haven, CT and London: Yale University Press, 2007), pp. xv-xviii.

5 J. North, *Cosmos: An Illustrated History of Astronomy* (Chicago, IL and London: University of Chicago Press, 2008), p. 417.

in London (1931). Contentious from the beginning, it proved to be highly influential.[6] It gave the study of Newton's life and work a twist which eventually brought to light a Janus-like Newton, engaged in 'scientific' and 'unscientific' pursuits. This apparent contradiction still provides food to be digested by the 'Newton industry'.

The paper effectively launched the persistent debate regarding 'externalism' and 'internalism' in the history of science. As it turned out, a young historically- and philosophically-minded biochemist attended the Congress and became a controversial participant in the debate. He was Joseph Needham, who forty years later re-affirmed his allegiance to Hessen's seminal ideas as follows:

> The trumpet-blast of Hessen may ... still have great value in orienting the minds of younger scholars towards a direction fruitful for historical analyses still to come, and may lead in the end to a deeper understanding of the mainsprings and hindrances of science in East and West, far more subtle and sophisticated than he himself could ever hope to be.[7]

Needham's two 'Grand Questions'

To paraphrase Cunningham and Williams: like it or not, a big picture of the history of science cannot avoid Needham's two 'Grand Questions'. This is not to say that they ignore Needham's 'Grand Project', *Science and Civilisation in China* (1954-), but that they err in ascribing to him, for example, the reductionist view 'that science *was* [sic] human civilization'.[8] Needham's opposition to gross reductionism emerges from an autobiographical piece (1972) in which he clearly declares that, after many intellectual struggles, he has reached the conviction that life consists in several

> irreducible forms or modes of experience. One could distinguish the philosophical or metaphysical form, the scientific form, the historical form, the aesthetic form and the religious form, each being irreducible to any of the

---

6   B. Hessen, 'The Social and Economic Roots of Newton's "Principia"', in *Science at the Cross Roads*, 2nd ed. (London: Cass, 1971), pp. 151-212. For relatively recent re-appraisals of Hessen and the Congress, see C. A. J. Chilvers, 'The Dilemmas of Seditious Men: The Crowther-Hessen Correspondence in the 1930s', *The British Journal for the History of Science*, 36 (2003), 417-35; idem, 'La signification historique de Boris Hessen', in S. Gerout (ed.), *Les Racines sociales et économiques des Principia de Newton* (Paris: Vuibert, 2006), pp. 179-206; G. Freudenthal, 'The Hessen–Grossman Thesis: An Attempt at Rehabilitation', *Perspectives on Science*, 13/2 (Summer 2005), 166-93.
7   Needham, *Science at the Cross Roads*, p. ix.
8   Cunningham and Williams, 'De-centring', p. 412.

others, but all being interpretable by each other though sometimes in flatly contradictory ways.[9]

And a year later, he states no less clearly:

> For myself I want to say I have not lost faith in science as a *part* [MT] of the highest civilization, and in its development as one single epic story for the whole of mankind.[10]

While the Grand Project has been widely acclaimed, Needham's approach has been criticised if not rejected. Both the relevance of the historical issue that first motivated Needham and the way he proposed to deal with it have come under fire. In the course of his work, Needham discovered that actually there were two questions to be answered. This is what he wrote, a quarter of a century or so after he formed the idea of what became *Science and Civilisation in China*:

> I regarded the essential problem as that of why *modern* science (as we know it since the +17th century, the time of Galileo) had not developed in Chinese civilization (or in Indian) but only in Europe. As the years went by, and as I began to find out something at last about Chinese science and society, I came to realize that there is a second question at least equally important, namely why, between the -1st century and the +15th century, Chinese civilization was much *more* efficient than occidental in applying human natural knowledge to practical human needs?
>
> The answer to all such questions lies, I now believe, *primarily* [MT] in the social, intellectual, and economic structures of the different civilizations.[11]

---

9 H. Holorenshaw [J. Needham], 'The Making of an Honorary Taoist', in M. Teich and R. Young (eds.), *Changing Perspectives in the History of Science* (London: Heinemann Educational Books, 1973), p. 7.

10 'An Eastern Perspective on Western Anti-science (1974)', in J. Needham, *Moulds of Understanding a Pattern of Natural Philosophy*, ed. by G. Werskey (London: Allen and Unwin, 1976), p. 297.

11 J. Needham, 'Science and Society in East and West', in M. Goldsmith and A. Mackay (eds.), *The Science of Science Society in Technological Age* (London: Souvenir Press, 1964), pp. 127-28. I am referring to this publication not because it originated as a letter to me, but because it contains what is regarded as Needham's 'classical' formulation of his Grand Question. Indeed the German version, 'Wissenschaft und Gesellschaft in Ost und West', heads the collection that Needham sanctioned of his own essays: J. Needham, *Wissenschaftlicher Universalismus Über Bedeutung und Besonderheit der chinesischen Wissenschaft*, ed. and trans. by T. Spengler (Frankfurt am Main: Suhrkamp, 1979), pp. 61-86. Later Needham re-iterates that there was not one question but two: 'Not only why modern science originated in Europe alone, but why, during the previous fifteen centuries, China had been much more advanced in science and technology than the cultures of the West'. See 'Man and His Situation (1970)', in Needham, *Moulds*, p. 282.

Here, for the purpose of this book, we must consider why Needham's case for the Scientific Revolution happening in Europe, and not in China, should not be excluded from a serious debate about the Revolution's make-up. The provision of chapter and verse may stand as an excuse for the number and length of quotations. Loose censures and generalised antagonistic assertions, born of politics and ideology, should not pass unnoticed either.

Nathan Sivin

Among the critiques of Needham's approach, Nathan Sivin's essay 'Why the Scientific Revolution Did Not Take Place in China – or Didn't It?' has attracted particular attention for understandable reasons.[12] Sivin had worked with Needham, who thought of him very highly, calling him 'a brilliant investigator of medieval Chinese astronomy and alchemy'.[13]

As far as I can make out, Needham's name appears in the text twice. First, in the introductory paragraph in which it is stated that 'Joseph Needham has given the "Scientific Revolution problem" its classic formulation'.[14] Second, in the fourth section of the essay (Fallacies), in the context of Needham's discussion of the *Book of Changes*, containing a classified system of symbols explaining all natural processes. By and large, Sivin prefers to refer to Needham in the notes. 'Because of his knowledge of the Chinese sciences and the breadth of his hypotheses', we read, 'Needham's is the earliest discussion of the Scientific Revolution problem that still commands attention, and is still the best'.[15]

This comment is in stark contrast to what we encounter in the text. True, Needham is not named but can there be doubt that his approach is the target of unmistakable disapproval if not rejection:[16]

> ...Why didn't the Chinese beat Europeans to the Scientific Revolution? – happens to be one of the few questions that people often ask publicly about why something didn't happen in history. It is analogous to the question of why your name did not appear on page 3 of today's newspaper. It belongs to an infinite set of questions that historians don't organize research programs around because they have no direct answers.

---

12 N. Sivin, 'Why the Scientific Revolution Did Not Take Place in China – or Didn't It?', in E. Mendelsohn (ed.), *Transformation and Tradition in the Sciences: Essays in Honor of I. Bernard Cohen* (Cambridge: Cambridge University Press, 1984), pp. 531-54.
13 Needham, *Moulds*, p. 283.
14 Sivin, 'Why', p. 531.
15 Ibid., p. 552, n. 7.
16 For the following paragraphs, see ibid., pp. 536-37, 539, 540, 543.

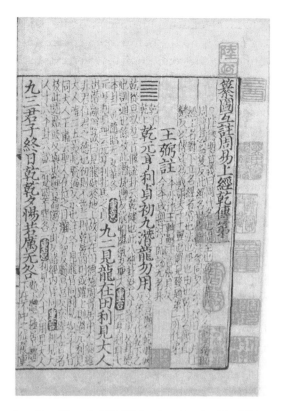

Fig. 12 A page from *Song Dynasty* (960-1279), printed book of the *I Ching (Yi Jing), Classic of Changes* or *Book of Changes*. National Central Library, Taipei City, Taiwan.

While the object is not explicitly identified in the text, the essay constitutes a massive assault on Needham's approach to the history of science and technology in China. Indeed, authored by a leading scholar in the field, it is frequently cited as a valid critique of Needham's answer to his two questions.

The point here is not to dispute Sivin's unquestionable sinological expertise. What is open to discussion is his belief that the distinction between external and internal factors has no objective reality but exists only as something perceived by historians subjectively.

I call it into question in the light of my long-standing interest in the history of fermentation, possibly the earliest biochemical process used by man in daily life. What has emerged from the historical study of the phenomena of fermentation is that they belong as much to the history of biochemistry (and other sciences) as to the technological and economic history of several industries. Another feature of the history of fermentation is that it was intimately connected with endeavours to answer the perennial question of the origin of living matter and the nature of the processes underlying it. A full historical treatment of the connection between extra-scientific and intra-scientific influences on fermentation theory and practice would require, for example, a consideration of patriotism. National feeling certainly prompted L. Pasteur's (1822-1895) work on beer after France's defeat by Prussia in the Franco-Prussian War (1870-1871).

This does not mean that the pursuit of 'external' and 'internal' aspects of the history of sciences respectively is meaningless. Both are justified provided we understand that the questions asked and the answers given are, by their very nature, partial and limited. The point at issue for an historian is how to amalgamate them, because history is in reality an integral process, not divided into 'external' and 'internal' compartments, but governed by a multitude of circumstances deriving from and belonging inseparably to spheres 'outside' and 'inside' science.[17]

The line taken by Sivin conforms to Steven Shapin's cogent observation regarding the externalism-internalism issue: 'Instead of clarifying and engaging with the problematic, I think over the last ten or fifteen years we rather turned our backs on it and wished it away'.[18]

A. C. Graham

The sinologist A. C. Graham also queries the relevance of asking why an event such as the Scientific Revolution did not happen in China. While expressing highest admiration for Needham, Graham cannot resist revealing that his proficiency in the Chinese language is not up to par.[19]

Graham's target is Needham's original historiographical sin. That is, Graham's assumption that the absence or presence of one set of conditions – the rise of capitalism – underlies the non-genesis or genesis of modern science in seventeenth-century China and Europe respectively. Out of the

---

17 M. Teich with Dorothy M. Needham, *A Documentary History of Biochemistry, 1770-1940* (Leicester and London: Leicester University Press, 1992), pp. xxii-xxv.

18 S. Shapin, 'Discipline and Bounding: The History and Sociology of Science Seen through the Externalism-Internalism Debate', *History of Science*, 30 (1992), 333-69 (p. 334).

19 A. C. Graham, 'China, Europe, and the Origins of Modern Science: Needham's The Grand Titration', in Nakayama and Sivin (eds.), *Chinese Science: Explorations of an Ancient Tradition* (Cambridge, MA: MIT Press, 1973), p. 46, n. 2.

many features of capitalism, Graham chooses capitalist profit as 'the only conscious goal that anyone has been able to find in the social processes that led to modern science'.[20] Whatever historiographical errors Needham may have committed, the absence of capitalist profit was not central to his analysis of factors that inhibited the rise of modern science in China. Graham's version of Needham's general emphasis on the social and economic factors greatly simplifies the latter's approach to history.

Sivin and Graham's reasonable starting point is that historians have enough on their plate without focusing on events that did not even happen. Nevertheless, as the great historian Eric Hobsbawm observes:

> *Conjectural* history has a place in our discipline, even though its chief value is to help us assess the possibilities of present and future, rather than the past, where its place is taken by *comparative* history; but actual history is what we must explain. The possible development or non-development of capitalism in imperial China is relevant to us only insofar as it helps to explain the actual fact that this type of economy developed fully, at least to begin with, in one and only one region of the world. This in turn may be usefully contrasted (again in the light of general models) with the tendency for other systems of social relations – for example the broadly feudal – to develop much more frequently and in a greater number of areas. The history of society is thus a collaboration between general models of social structure and change and the specific set of phenomena which actually occurred. This is true whatever the geographical or chronological scale of our enquiries.[21]

Using conventional intellectual criteria, Sivin suggests China had its own scientific revolution (lower case!) in the seventeenth century. This is intriguing on three accounts. First, Sivin bases his suggestion on the same kind of comparison as Needham, who is criticised for expecting China to follow the same route as Europe. Second, the term scientific revolution/Scientific Revolution is commonly associated with radical transformation in the human perception of the motions of celestial *and* terrestrial bodies in the context of a unified theory. But Sivin limits the meaning of the term to changes amounting to a conceptual revolution in *astronomy*. Last but not least, Sivin still admits:

> That revolution did not generate the same pitch of tension as the one going on in Europe at the same time. It did not burst forth in as fundamental a

20 Ibid., p. 67.
21 Hobsbawm, *On History*, p. 80.

reorientation of thought about Nature. It did not cast doubt on *all* the traditional ideas of what constitutes an astronomical problem, and what significance astronomical prediction can have for the ultimate understanding of Nature and of man's relation to it.

Most important, it did not extend the domain of number and measure in astronomy until it embraced every terrestrial phenomenon (the Jesuits were obliged to conceal from the Chinese the development in Europe). What happened in China bears comparison with the conservative revolution of Copernicus rather than with the radical mathematization of hypotheses Galileo precipitated.[22]

Consequently, I would say the two movements are different beasts.

H. Floris Cohen

It is remarkable how authors who go out of their way to sing Needham's praises are also at pains to demonstrate flaws in his approach to the history of science. A telling exemplar is H. Floris Cohen's ambivalent treatment of Needham's answer to his own Grand Question. To this theme, Cohen devotes seventy pages (out of 662) in a historiography of the Scientific Revolution stimulating not least for its combative style. Combative in the sense that, while wishing to be fair and giving his due to scholars from A(iton) to Z(ilsel), Cohen is unsparing with censures right and left. This gives the impression that Cohen has a viewpoint on Needham and the Grand Question. Regrettably, that is not the case: what he offers is 'a highly provisional sketch' in the last chapter of the book:

> *Has the time come for answering the Grand Question?* I would not so easily bring up the question just put, which I am in the end not qualified to answer, if it had not in effect been asked by some of Needham's most knowledgeable critics. Although they have not said so in quite so many words, it has been suggested by Nakayama and Sivin in particular (and much in our preceding account tends to support such a suggestion) that just about everything Needham has ever written to answer his Grand Question should be regarded as *one giant projection of Joseph Needham's collected preconceptions upon China's society and world of thought.*[23]

---

22 Sivin, 'Why', p. 540.
23 See H. F. Cohen, *The Scientific Revolution: A Historiographical Inquiry* (Chicago, IL and London: University of Chicago Press, 1994), p. 471.

When all is said and done what separates Needham from Sivin, Graham and other critics (on whom Cohen draws) is the weight he gives to the part played by social and economic factors in the generation of the Scientific Revolution in Europe:

> Whatever the ideological inhibiting factors in the Chinese thought-world may turn out to have been, the certainty always remains that the specific social and economic features of traditional China are connected with them. They were clearly part of that particular pattern, and in these matters one always has to think in terms of a package-deal. In just the same way, of course, it is impossible to separate the scientific achievements of the Ancient Greeks from the fact that they developed in a mercantile, maritime, city-state democracy … It may be that while ideological, philosophical and theological differences are never to be undervalued, what mattered most of all were the facilitating pressures of the transition from feudalism to mercantile and industrial capitalism, pressures which did not effectively operate in any culture other than that of Western, Frankish Europe.[24]

Robert Finlay

Last but not least, let us turn to Robert Finlay's portrayal of Needham's perception of the Scientific Revolution. That emerges in the context of Finlay's presentation of Needham's perspective on the celebrated sea-voyages that took place during the early years of the Ming dynasty (1368-1644).[25]

Between 1405 and 1433, seven vast fleets sailed into the Indian Ocean and reached the eastern shores of Africa. Each fleet comprised over 200 sailing ships with about 27,000 men on board. Some of these vessels were the longest wooden ships ever built (117-134m long).[26]

---

24 Needham, *Moulds*, pp. 285-87.
25 R. Finlay, 'China, the West, and World History in Joseph Needham's Science and Civilisation in China', *Journal of World History*, 11 (2000), 265-303. See J. Needham with Wang Ling and Lu Gwei-Djen, *Science and Civilisation in China*, Vol. 4/3: 'Civil Engineering and Nautics' (Cambridge: Cambridge University Press, 1971), pp. 487-535. For a convenient roundup (also referred to by Finlay), see J. Needham, 'Abstract of Material Presented to the International Maritime History Commission at Beirut', in M. Mollat (ed.), *Sociétés et Compagnies de Commerce en Orient et dans L'Océan Indien* (Paris: S.E.V.P.E.N, 1970), pp. 139-65 (pp. 146-54).
26 For an acclaimed account, see E. L. Dreyer, *Zheng He: China and the Oceans in the Early Ming Dynasty* (New York and London: Pearson Longman, 2007). 'It will be', so the knowledgeable Jonathan Mirsky claims, 'the last word for some time to come'. See his review 'Tribute, Trade and Some Eunuchs', *The Times Literary Supplement*, 26 January 2007, 11.

Fig. 13 Zheng He's Treasure Ship. Model at the Hong Kong Science Museum.

Both the purpose of the state-directed voyages, under the command of the eunuch Zheng He (Chêng Ho) (1371-1433), and the reason why they were terminated have been much debated in the West. Indeed, Finlay states, they are

> now renowned in the West precisely because Needham made them known to a wide audience for the first time. His dramatic contrast of these Ming expeditions with those of the Portuguese in the early sixteenth century captured the imagination of his readers. Rarely has a historical study been so widely applauded and universally accepted. It has been integrated into influential interpretations of world history… But the context which Needham establishes for the voyages and the part they play in his conception of world history remain unrecognized.[27]

Finlay's intent is to put the record straight. Needham's primary offence is that '[he] regards the "proto-scientific" motive as most significant … stemming from his determination to present the Ming expeditions as embodying the virtues of China in contrast to the vices of the West'.[28]

Finlay's claim is not supported by the list of motives that Needham names one by one. There the pursuit of natural and medical knowledge occupies fifth place among seven.[29]

In fact, Needham's view of the purpose of the voyages, stated in 1970, accords with Mirsky's summary of the 'state-of-the-art' 37 years later:

---

27 Finlay, 'China', 269.
28 Ibid., 293, 299.
29 Needham, 'Abstract', 147.

While it is true that Zheng He brought back, for the exclusive use of the Ming household, lions, leopards, ostriches and giraffes, spices and minerals – his largest vessels were called Treasure Ships – the purpose of the voyages, as the main authorities now agree, was to enfold distant rulers, some of whom sent their envoys to China on Zheng He's ships, in the ancient Chinese 'tribute system'. According to this tradition the Emperor, ruling from the centre of the world, by his virtue and splendour attracted foreigners to his Court. There they presented him with their goods, deemed to be 'tribute' and performed the kowtow.[30]

Finlay's flawed presentation of Needham's position on the purposes of the voyages is not an isolated error. His account abounds with inaccuracies. Some are trivial, such as blaming Needham for casting Zheng He as the 'Vasco da Gama' of China. In fact, Needham attributes this designation to Frank Debenham, the author of *Discovery and Exploration: An Atlas of Man's Journey into the Unknown* (1960). More serious is Finlay's assertion that '*Science and Civilisation* is a product of the Cold War', conceived 'when the People's Republic was proclaimed'. Oddly, Finlay contradicts himself because a few pages previously he writes:

> Liberated by his exposure to Chinese culture in 1937, Needham first conceived of writing a book on Chinese science and technology in 1942, after completing *Biochemistry and Morphogenesis* and before leaving for China.[31]

As it happens, we can date more precisely Needham's announcement, in the essay 'On Science and Social Change', of his intention to investigate after the war the failure of science

> to arise in China [which is] one of the greatest problems in the history of civilization, since no other culture, accept the Indian, equals that of the West in scope and intricacy more closely than that of China.[32]

This essay was composed in September 1944, when Needham found himself

> in a vacuum within a vacuum, cut off from all communication with the outside world by landslides and rockfalls, without telegraph or telephone, awaiting the clearing of the road.[33]

---

30 Mirsky, 'Tribute'.
31 Finlay, 'China', 301, 273-74.
32 'On Science and Social Change', in J. Needham, *The Grand Titration: Science and Society in East and West* (London: Allen & Unwin, 1969), pp. 123-54 (p. 148). The essay first appeared in *Science and Society*, 10 (1946), 225-51.
33 Ibid., p. 123.

Ostensibly written for a collection on social organisation, the essay is about many things: science, philosophy, religion, social orders, their history in the West and in China. More specifically, it is about the intrinsic incompatibility of science with Nazism and Shinto-fascism because of the inherently democratic nature of science. Needham then describes democracy as 'that practice of which science is the theory'.[34]

Finlay refers to this essay several times without due care to chronology. That is, without noting that Needham was *then* clearing his mind for the composition of *Science and Civilisation*, while waiting for the clearing of the road to Burma. The theme of the correlation between science and democracy had surfaced in Needham's previous historical work, including *The Levellers and the English Revolution* (1939), published under the pseudonym Henry Holorenshaw (with a foreword by himself).[35] Along the lines discussed above, Needham observes in 1944 that the rise of the merchant class to power, with their slogan of democracy, was the indispensable accompaniment and *sine qua non* of the rise of modern science in the West. But in China the scholar-gentry and their bureaucratic feudal system always effectively prevented the rise to power or seizure of the State by the merchant class, as happened elsewhere.[36]

This approach is anathema to Finlay. He worries that

> Needham's conception of world history has gone unnoticed, however, and most readers certainly are not aware of a large-scale scheme informing the massive detail of *Science and Civilisation*.[37]

In the unremitting quest to alert and enlighten the ignorant reader, Finlay, like other critics, damns with faint praise.[38]

Finlay and Needham have different takes on China, the West and world history. But that does not justify him in misrepresenting Needham's view

---

34 Ibid., p. 145.
35 I concur with Finlay that there was no intention to deceive by presenting 'Henry Holorenshaw' as the author of 'The Making of an Honorary Taoist', in Teich and Young (eds.), *Changing Perspectives in the History of Science*. Originally, the publishers expected me to write on Needham. After considering the idea, it occurred to me that a contribution by Needham's *alter ego* would be of greater interest. Finding the suggestion intriguing, Needham accepted the invitation to contribute.
36 Needham, *Titration*, p. 150.
37 Finlay, 'China', 267.
38 For example: 'Even though he employs many outdated concepts and makes countless unsupported assertions, his rendering of world history is remarkable for its synoptic vision'; 'Although not without internal coherence and a grain of truth, his depiction of European history amounts to little more than a mechanical application of Marxist clichés that were outdated before the first volume of *Science and Civilisation* appeared'. Ibid., 268, 301.

of history. Needham believed that the rise of modern science was connected with the rise of mercantile capitalism in Europe, constituting an aspect of the transition from feudalism to capitalism. This does not warrant Finlay's attribution to Needham of the reductive worldview that

> the rise of modern science in Europe was not merely the outcome of developments within Western civilization but was a direct result of the diffusion of Chinese technology; hence, if China did not give birth to the Scientific Revolution, it at least was responsible for its conception. In other words, the transition to modernity sprang from the long-distance influence of China on the West.[39]

It is noteworthy that Finlay refrains from citing any of Needham's writings as support for this statement – although he refers to them copiously.

Chinese society, agriculture and technical progress

Needham's accent on the divergence of the Scientific Revolution in Europe and China connects with broader questions.

It was due to Marxian ideas that Needham regarded the Scientific Revolution as an aspect of the transition from feudalism to capitalism in Europe. He held the fact that 'the development of modern science occurred in Europe and nowhere else' to be part and parcel of European historical singularity.[40] This perspective has been recently addressed by the distinguished

---

39 Ibid., 281-82.
40 Needham, *Titration*, p. 193. For a recent thoughtful reappraisal of Needham's 'settled/outdated' position, see P. K. O'Brien, 'The Needham Question Updated: A Historiographical Survey and Elaboration', *History of Technology*, 29 (2009), 7-28. O'Brien finds Needham's point intact about the divergent attitudes of European and Chinese natural philosophy to 'laws of nature'. At the top of the agenda for historical research, he concludes,

> must be the Chinese stance of incredulity towards the paradigm that had gripped the imagination of European natural philosophy, namely that all natural phenomena, including the human body, could be investigated, comprehended and interrogated as cases or instances of universal laws of nature. Furthermore, these laws (which explained how and why things operated as they did) were the manifestations of the intelligent design of a divine creator. They could be exposed by transparent experimental methods and explicated rigorously in mathematical language. Natural laws that could be represented as divine in origin provided the West with a cosmology and a culture for elites of aristocrats, merchants, industrialists and craftsmen that rested on an acceptable, unproveable, but ultimately progressive supposition that God created a natural world that was rational and explicable, that its tendencies to afflict the lines of people's everywhere could be fixed or ameliorated and that matter could be manipulated to provide technologies to raise the productivities of labour.

Austrian social historian Michael Mitterauer, who points to contemporaries of Max Weber (his great inspiration) acknowledging as 'valid' science only the one developed in the West.[41]

Mitterauer does not concern himself with either the Scientific Revolution or the transition from feudalism to capitalism. Nevertheless, what emerges from his account is that a string of factors operating within Frankish feudalism (agriculture and technical innovations, manorial/estate system, commerce, crusades, printing) had generated, by the fourteenth and fifteenth centuries, circumstances conducive to the rise of capitalism in some Italian and Flemish cities. Pushed forward by the incipient world market economy, these factors may be (following Marx) identified: growth of trade, employment of free labour in mining and manufacturing, separation of the peasantry from the land, and coercive accumulation of surpluses ('primitive accumulation of capital').

Medieval China did not participate in the burgeoning world market economy, associated with the global expansion of trade by European powers.[42] This had something to do with its underdeveloped money economy and the non-maturation of its capitalist relations, two sides of the same coin. Moreover, the size of the country and its population put a premium on self-sufficiency in food production, demanding both extension of the cultivated area and intensive agriculture; these, in turn, required more labour. This is where, in Needham's terms, the 'bureaucratic feudal state system' distinguished itself by not driving the agricultural producers from land. The intensification of agriculture doubtless stimulated inventions, such as the water-powered

---

Confucian cosmology neither restrained nor promoted the interrogation of nature or the search for technological solutions to problems of production. What it did not provide for, even during the continued economic advance of the Qing empire, was that powerful promotional confidence that entered into the cultures of Western elites of a natural world that was the rational and explicable work of their God. As Needham observed, 'there was no confidence that the codes of nature could be read because there was no assurance that a divine being had formulated a code capable of being read'. His point is intact and remains open for research and discussion.

Ibid., 28. The Epilogue will further discuss the conundrum of European natural philosophers faced with harmonising their findings and explanations of natural phenomena with the words of Scripture.

41 M. Mitterauer, *Why Europe? The Medieval Origin of its Special Path* (Chicago, IL and London: University of Chicago Press, 2010), p. xx.

42 What follows draws on my comments on the manuscript of the public lecture by J. Needham at the University Hong Kong on 30 April 1974. See J. Needham and R. Huang (Huang Jen-Yü), 'The Nature of Chinese Society – A Technical Interpretation', *Journal of Oriental Studies*, reprinted from Vol. 12/1 and 2 (1974), 1-16. See also J. Needham, with Wang Ling, *Science and Civilisation in China*, Vol. 4/2 (Cambridge: Cambridge University Press, 1965), p. 262.

blowing engine for the casting of iron agricultural implements. The tendency to rely on self-sufficiency encouraged technical developments, but it also derived momentum from them.

Before the impact of microelectronics, it can be said that machinery essentially consisted of three elements: the driving, the transmission and the actual operational mechanism. In the light of this assortment, the characteristic Chinese achievements, such as the gear-wheel, the crank, the piston-rod and the method of inter-conversion of rotary and longitudinal motion, belong to the category of the transmission mechanism. Revolutionary changes in production, in the European context of the transition from feudalism to capitalism at any rate, appear to have been connected with the working element of the machinery, which had replaced manual operations in the production processes. The flying shuttle and the spinning jenny preceded Watt's steam engine, but they were hand-powered.

Chinese inventions, such as the humble wheel barrow, were labour saving in an agrarian society which needed labour to produce food in order to satisfy the basic requirements of a very sizeable population; but they could hardly become the starting point of radical changes in production, since, paradoxically, no surplus labour existed. Given that in China the monetary system was underdeveloped, the Chinese bureaucratic feudal state organisation collected taxes and paid its apparatus in kind, but proved to be incompatible with the creation of surplus capital. This then reinforced the immobility of the agrarian society, hindered its social stratification, and, despite the existence of a tremendous reservoir of technical skill, retarded the stimuli to the commercialisation of agriculture and the development of productive forces.

6. West and East European Contexts

The demise of the feudal system in Europe: technical development and international trade

Printing, gunpowder and the magnetic compass have long been cited as hallmarks of the cultural and technical superiority of medieval China. This has been the case ever since Francis Bacon pinpointed them as the three discoveries, unknown to the ancients, that

> have changed the whole face and state of things throughout the world; the first in literature, the second in warfare, the third in navigation; whence have followed innumerable changes, insomuch that no empire, no sect, no star seems to have exerted greater power and influence in human affairs than these mechanical discoveries.[1]

Rupert Hall, writing some 350 years later, confirms Bacon's opinion on the marked effects of the discoveries of gunpowder and printing on European society:

> We might well choose to date the beginning of modern European history from the introduction of gunpowder. By 1325 primitive cannon were in action, and from 1370 mechanical artillery (on the lever principle) was falling into abeyance. By 1450 the hand gun had appeared, beginning the obsolescence of crossbow and longbow. Powder-making became an important industry, along with cannon-founding and gun-making. By 1500 heavy guns, mortars, and explosive mines had made the medieval castle almost untenable ...

---

1   F. Bacon, 'Aphorisms – Book One', in *The New Organon and Related Writings*, ed. by F. H. Anderson (New York and London: MacMillan, 1987), p. 118.

http://dx.doi.org/10.11647/OBP.0054.06

Only the wealthiest and most powerful rulers could afford the new gunpowder weapons and the more expensive and larger armies. The rebellious feudal nobility in their isolated castle fortresses could no longer withstand the power of the monarch's more powerful gunpowder weapons. Thus the changes in the technology of warfare brought about through the introduction of gunpowder aided in that process of administrative and territorial consolidation which was to give rise to monarchical states and to the nation-state system of Europe as one knows it today ... Just as the invention of gunpowder served to batter down the political positions of feudalism, so did the invention of printing help to remove the barriers of communication between men. And, like gunpowder, the invention has had incalculable significance to human history, far beyond its immediate technological effect.[2]

For the historian the problem is that the use of gunpowder and printing in the wake of technological improvements (incendiary properties of mixtures of nitre, sulphur and charcoal; papermaking; printing with movable type) became widespread in Europe but not in China before 1600. How and why this came about has been the subject of continuous examination, with pride of place rightfully assigned to Arabic-Islamic mediation.

In this context Hall's periodisation of the beginning of modern European history is noteworthy. He associates it with the demise of the feudal system in Europe – in which the spread of gunpowder and movable type printing played a significant part. Without spelling it out, Hall clearly implies that a societal transformation was at play, conceptualised by Marxist historians as a period of transition from feudalism to capitalism. Though doubtless Hall, a stern critic of the Marxist approach to history, would have been horrified to be aligned with it.

Be that as it may, research bears out the historical role of merchant capital (discussed earlier) in undermining feudalism in Europe through international trade. Moreover, research has found it relevant to the understanding of the genesis of the Scientific Revolution. Thus, in the concluding chapter of a penetrating study of commerce, medicine and science in the Dutch Golden Age (c. 1550-1700), Harold J. Cook has this to say:

---

2   A. Rupert Hall, 'Early Modern Technology to 1600', in M. Kranzberg and C. W. Pursell, Jr. (eds.), *Technology in Western Civilization*, Vol. 1 (New York: Oxford University Press, 1967), pp. 98-100. There is, of course, also the question of the stagnation that affected Islamic civilisation. Here Michael Mitterauer's discussion of Muslim hostility to the printing of the Qur'an is illuminating. Thus, the first printing-house in Istanbul to work with Arabic letters was opened in 1726 only to be shut down from 1730 to 1780 and again in 1800. See M. Mitterauer, *Why Europe? The Medieval Origin of its Special Path* (Chicago, IL and London: University of Chicago Press, 2010), p. 266.

It was no accident, then, that the so-called Scientific Revolution occurred at the same time as the development of the first global economy. That world linked the silver mines of Peru to China as well as Europe, the sugar plantations of the Caribbean and the nutmeg-growing regions of Southeast Asia to slave labor as well as to luxury consumer goods, and a wealth of new information circulating in coffeehouses and lecture halls to books and the natural objects available in European gardens, cabinets of curiosity, and anatomy theaters.[3]

Pursuits of natural knowledge: The Court of Rudolf II in Prague

Among European gardens and cabinets, the ones created by Emperor Rudolf II in Prague acquired particular fame. As the knowledgeable Paula Findlen puts it:

> Rudolf assembled a rich array of flora and fauna and an extensive collection of specialized instruments that might help him pursue knowledge of nature. At the height of his reign, a deer park surrounded Hradčany Castle, complemented by an aviary (in which one might glimpse the Emperor's birds of paradise and, after 1598, a dodo) and a botanical garden where distinguished naturalists such as Rembert Dodoens and Carolus Clusius tended exotic plants. From the 1580s onwards, a steady stream of visitors such as the English alchemists John Dee and Edward Kelley, the Italian mystic and Neoplatonist Giordano Bruno, and the physicians and occultists Oswald Croll and Michael Maier enjoyed audiences with Rudolf, bearing gifts of magic talismans, their own writings and promises to unlock the secrets of nature.[4]

---

3 H. J. Cook, *Matters of Exchange: Commerce, Medicine, and Science in the Dutch Golden Age* (New Haven, CT and London: Yale University Press, 2007), p. 411. Intentionally or not this passage echoes Marx's analysis of what he characterised as 'primitive accumulation of capital':

> The discovery of gold and silver in America, the extirpation, enslavement and entombment in mines of the aboriginal population, the beginning of the conquest and looting of the East Indies, the turning of Africa into a warren for the commercial hunting of black-skins signalized the rosy dawn of the era of capitalist production. These idyllic proceedings are the chief momenta of primitive accumulation. On their heels treads the commercial war of the European nations, with the globe for a theatre. It begins with the revolt of the Netherlands from Spain, assumes giant dimensions in England's anti-Jacobin war, and is still going on in the opium wars against China.

See K. Marx, *Capital*, Vol. 1 (London: George Allen & Unwin, 1938), p. 775.

4 P. Findlen, 'Cabinets, Collecting and Natural Philosophy', in E. Fučíková et al., *Rudolf II and Prague* (Prague: Prague Castle Administration and London and Milan: Thames and Hudson, 1997), pp. 213-14. The volume provides a notable introduction in English to Rudolf II and his era. See also R. J. W. Evans, 'Rudolf II: Prag und Europa um 1600', in *Prag um 1600 Kunst und Kultur am Hofe Kaiser Rudolf II*, Vol. 1 (Freren: Luca Verlag, 1988), pp. 27-37. For Evans's previous full-length treatment, see *Rudolf II and His World:*

These persons do not exhaust the circle of prominent pursuers of magic, mysticism, alchemy and knowledge of nature who were connected with Rudolf's court. Under Rudolf's auspices and at the behest of Hagecius (mentioned earlier), the collaboration of Brahe and Kepler came about. Eventually, this partnership was to transform Copernican astronomy into genuine heliocentrism. Equipped with mathematical and astronomical skills, the excellent instrument makers Jo(o)st Bürgi and Erasmus Habermehl were at hand to help.[5] The Imperial mathematician Nicholas Reymarus (Raymers) Baer (Ursus) developed a kind of heliocentric system, one that famously prompted Brahe to charge him with plagiarism. As North pointed out: 'Baer's system differed from Tycho's in one important respect: he gave the earth a daily rotation on its own axis, half-loosening it from its old bonds, so to speak'.[6]

Rudolf's patronage of Brahe and Kepler has long been known. What historians of astronomy seemed not to have noticed, before the Czech scholar Josef Smolka spotted it, is a remarkable passage in Kepler's letter to Galileo on 8 April 1610 (Julian calendar). It records that Rudolf carried out his own astronomical observations:

A Study in Intellectual History, 1576-1612 (Oxford: Clarendon Press, 1973). For a Czech historiographical perspective, see J. Válka, 'Rudolfine Culture', in M. Teich (ed.), *Bohemia in History* (Cambridge: Cambridge University Press, 1998), pp. 117-42. Here I cannot but note the recent publication of the massive multi-authored survey of alchemical activities associated with Rudolf II's court: see I. Purš and V. Karpenko (eds.), *Alchymie a Rudolf II: Hledání tajemství přírody ve střední Evropě v 16. a 17. století* [*Alchemy and Rudolf II: Searching for Secrets of Nature in Central Europe in the 16th and 17th Centuries*] (Prague: Artefactum Ústav dějin umění AV ČR, 2011).

5   A. Švejda, 'Prager Konstrukteure wissenschaftlicher Instrumente und ihre Werke', in J. Folta (ed.), 'Science and Technology in Rudolfinian Time', *Acta historiae rerum naturalium necnon technicarum. Prague Studies in the History of Science and Technology*, Vol. 1 (1997), 90-4.

6   J. North, *Cosmos: An Illustrated History of Astronomy* (Chicago, IL and London: University of Chicago Press, 2008), p. 336. For an analysis of the Baer-Brahe controversy, see N. Jardine, *The Birth of History and Philosophy of Science: Kepler's 'A Defence of Tycho against Ursus' with Essays on its Provenance and Significance* (Cambridge: Cambridge University Press, 1988). For a scholarly, bilingual French-Latin collection of related documents, see N. Jardine and A.-Ph. Segonds, *La Guerre des Astronomes: La Querelle au sujet de l'origine du système géo-héliocentrique à la fin du XVIe siècle*. Vol. 1: Introduction (Paris: Belles lettres, 2008); N. Jardine and A.-Ph. Segonds, *La Guerre des Astronomes: La Querelle au sujet de l'origine du système géo-héliocentrique à la fin du XVIe siècle*. Vol. 2/1: *Le 'Contra Ursum' de Jean Kepler, Introduction et textes préparatoires* and Vol. 2/2: *Le 'Contra Ursum' de Jean Kepler, Édition critique, traduction et notes* (Paris: Belles lettres, 2008). It has been suggested that Brahe's furious reaction had something to do with the differing social provenances of the protagonists. Whereas Brahe was of noble parentage, Baer was of peasant stock and hence less trustworthy.

It is three months that the Emperor put to me various questions regarding the constitution of moonspots and thought the images of lands and continents shine brightly on the Moon or in the telescope. He asserted to have probably seen the image of Italy with two nearby islands. In the next days he also placed the telescope at disposal for the same kind of observations but no use was made of it. At the same time, Galileo … through your beloved observations the Monarch of the Christian world is surpassed.[7]

Rudolf II was also a patron of artists, such as the Milanese painter Giuseppe Arcimboldo (1527-1593), whose portraits of human heads and paintings of elements and seasons effectively bridge art and natural history. They encompass detailed observations of flowers, fruit, vegetables and trees in the style identified as 'scientific naturalism'. One of its forerunners, if not founders, was Leonardo da Vinci, the other interdisciplinary Milanese artist whom Arcimboldo repeatedly wished to equal. As a recent commentator put it, Arcimboldo's paintings grow out of 'the mysterious interweaving of art, science and the occult in Renaissance Europe'.[8]

An early form of institutionalised science?

The pursuits of natural knowledge associated with the Prague court were by no means unique. Contemporarily, there were greater and lesser aristocratic courts in Europe suffused with astrological, alchemical and cabalistic thinking. Since the late 1950s, these activities have been discussed as an early form of the institutionalised pursuit of natural knowledge.

The idea was taken up and debated by a group of six historians of science, collaborating on a history of exact sciences in Bohemia (up to the end of the nineteenth century). In the published volume, the late and well-informed Zdeněk Horský refers to the 'Rudolfinian centre of scientific work' which 'was not a mere accidental result of the activity of foreign scientists temporarily staying at Prague but evolved with the direct participation of Thaddeus Hagecius in an environment in which astronomical research had been continually growing and deepening since the middle of the sixteenth

---

7 J. Smolka, 'Böhmen und die Annahme der Galileischen astronomischen Entdeckungen', *Acta historiae … Prague Studies in the History of Science and Technology*, Vol. 1 (1997), 41-69 (p. 49).
8 J. Jones, 'Natural Wonders', *Saturday Guardian*, 26 April 2008. See also T. DaCosta Kaufmann, *Arcimboldo: Visual Jokes, Natural History, and Still-Life Painting* (Chicago, IL: University of Chicago Press, 2009).

century'.[9] Horský reiterates this same position more than a quarter of a century later, albeit now referring to a 'scholarly centre' (*Gelehrtenzentrum*).[10]

Another knowledgeable member of the group, Josef Smolka, more than thirty years after their joint publication, voices justifiable scepticism about applying the term 'centre' to the circle of inquirers into natural phenomena supported by Rudolf II. It implies an organised and programmed activity, whereas 'we do not encounter anything of the sort with the exception of the short time Tycho's group was at large'.[11]

This has something to do with Rudolf's II apparent dislike of extreme Catholic confessionalism in spite of the Council of Trent's clarion call to Counter-Reformation (1545-1563). Not only were the Protestants Brahe and Kepler offered the opportunity to co-operate, but the Jew David Gans was able to strike up a relationship with both. A German Jew, educated in Poland and settled in Prague, Gans authored a variety of books on astronomy, mathematics, geometry, history and other topics. 'He did it in part', writes Noah J. Efron, 'because he was persuaded that these subjects might provide common language for Jews and Christians.'[12]

Fig. 14 David Gans, Ptolemaic cosmological diagram (planetary circles surrounded by Zodiac constellations) in Hebrew, from his *Nechmad V'Naim* (1743).

9   J. Folta et al., *Dějiny exaktních věd v českých zemích do konce 19. století* [*History of the Exact Sciences in the Czech Lands up to the End of the Nineteenth Century*] (Prague: Nakladatelství Československé akademie věd, 1961), p. 47.

10 Z. Horský, 'Die Wissenschaft am Hofe Rudolfs II', in *Prag um 1600*, pp. 69-74.

11 J. Smolka, 'Scientific Revolution in Bohemia', in R. Porter and M. Teich (eds.), *The Scientific Revolution in National Context* (Cambridge: Cambridge University Press, 1992), p. 220.

12 N. J. Efron, 'Liberal Arts, Eirenism and Jews in Rudolfine Prague', *Acta historiae... Prague Studies*, 24-35 (p. 24).

Conspicuously, the list of persons associated with Rudolf's court contains hardly any prominent local figures, apart from Hagecius. One was the astronomer Martin Bachacius (Bacháček) (1539/1541-1612), who collaborated with Kepler without apparently understanding the latter's heliocentrism. Bachacius is remembered less as an astronomer than as an organiser of higher education – he occupied the post of Rector of Protestant Prague University (*Collegium Caroli*, Karolinum) between 1598-1600 and 1603-1612.

Another Rector of Prague University, linked to the court, was born into a burgher family in Breslau/Wrócław. At Brahe's bidding, Jan (Johannes) Jessenius (1566-1620), a medical man with astronomical interests and diplomatic skills, came to Prague after performing a celebrated autopsy in 1600. He was among the 27 notables executed in the Prague Old Town Square (21 June 1621) for their leading role in the anti-Habsburg uprising of the Bohemian estates in 1618. Aristocrats by and large, their concern was to defend their power against the mounting centralisation of the Habsburg rulers. The 'Bohemian War', set in motion by the defenestration of the two highest state officials (and their secretary) from a window of the Prague Castle, is looked upon as the first phase of the Thirty Years War (1618-1648).

Robert Evans observes that there was a close connection between Rudolf II and the outbreak of the Thirty Years War. In 1609 he was forced to sign a famed 'Letter of Majesty', endorsing Czech Protestantism (the 'Czech Confession'). Accordingly, religious freedom was granted not only to nobles and knights, including their feudal subjects, but also to royal towns. But the then most liberal religious document in Europe did little to reduce tensions between the Catholic and Protestant creeds. However, there is more to the 'Bohemian War' and to the Thirty Years War than unresolved religious and political polarity.[13]

The Thirty Years War and the Scientific Revolution

Epitomised by the social and cultural life of the courts of Emperor Rudolf II and Queen Elizabeth I, Prague and London were cities of the same rank around 1600. During the next hundred years, however, the situation changed radically in the wake of interstate rivalry for a piece of the opening world market. It is in this context that the Thirty Years War and the Scientific

---

13 Evans, 'Rudolf II', p. 35.

Revolution became both products of and factors in the uneven transformation of late medieval (feudal) Europe to early modern (capitalist) Europe. And yet, remarkably, in traditional accounts of this period, we would be hard pressed to find explanations or even discussions of a link between two of its defining events. Hal Cook's exploration of the rise of medicine and science in the context of Dutch sea-bound commerce is a rare attempt to connect the scientific and the social dimension.

True, here and there, we read that in 1618 Descartes, an iconic figure in the story of the Scientific Revolution, joined the army of Maurice Nassau, the Protestant captain-general of the Dutch Republic, and later that of Count Tilly, the commander of the Catholic League. Stationed in the later part of 1619 at Ulm on the river Danube in Southern Germany, Descartes famously experienced a moment of revelation. According to his retrospective narrative, Descartes spent 'the whole day shut up in a stove-heated room' and 'at full liberty to discourse with myself about my own thoughts', when he began to develop his means of distinguishing truth from falsehood, a solution enshrined in the *Discourse on the Method* (1637). One wonders whether Descartes's longing to establish a criterion for truth had something to do with his service in the two opposed armies, each upholding the sole – Protestant/Catholic – religious truth. It should be noted that Descartes's service in the armies of both faiths became less exceptional as the war continued. Recruitment became a problem frequently solved by forcing prisoners of war to change sides.

The inclination to regard the Thirty Years War primarily in terms of German religious and political history has not declined. It connects with the fact that, geographically, the Holy Roman Empire of the German Nation became the arena of military operations. The limitation of this viewpoint becomes apparent when we consider the incursions of Danish and Swedish armies, the participation of France and Spain with money and soldiers, and the resistance of the Dutch Republic (at war with Spain) to the deployment of Spanish troops.

While the religious and political agenda cannot be passed over, it is not the limit of the Thirty Years War's place in history. This was a European war played out on over half the Continent, though not all parts were in action at the same time. It was also, in a sense, a global war: behind it lurked the rivalry of European states seeking a share in, if not domination of, the expanding world trade and world market. It outstripped previous military conflicts in material and financial costs, size of deployed armies, military and civil casualties and, last but not least, savagery.[14]

The need for money and credit provided the setting for intensified private and public money-lending activities, including money-changing. Such activities underpinned monetary circulation, a crucial feature of financial business in the pre-industrial phase of capitalism. But monetary transactions suffered from chaotic currency exchanges until some order was established in 1609, with the founding of the municipal Amsterdam Exchange Bank. Apart from fostering the trade of the northern Netherlands with the Baltic, the Levant and the Far East, the aim of the bank

> was the establishment of a clearing system based on transfers from one client's account to another's in bank guilders of account representing a constant silver content. These Amsterdam clearing transactions required a special dimension because of the city's dominant position in world trade at that time. The obligation imposed by the municipality to clear all bills of exchange of an amount above 600 guilders through the bank induced all-important Amsterdam merchants and in fact all important firms involved in world trade to open an account at the Amsterdam Exchange Bank, which thus grew in the course of the seventeenth century into a bank of world stature – indeed, into *the* great clearing house for international trade, with the stable bank guilder serving as the world's convertible key currency.[15]

Amsterdam's financial dominance was part and parcel of the long-drawn-out shift of commercial gravity from the Mediterranean to the Atlantic (c. 1200-1600). We can now see it as a phase in the history of world economic integration, one in which merchants from the northern and southern Low Countries were pre-eminently involved. A societal transformation was set in motion whereby merchant *burgers* not only metamorphosed into the capitalist bourgeoisie, but the Dutch also experienced the Scientific Revolution as examined by Cook in *Matters of Exchange*.

---

14 Texts dealing with the Thirty Years War and the period are legion. I draw attention to the undervalued J. V. Polišenský with F. Snider, *War and Society in Europe, 1618-1648* (Cambridge: Cambridge University Press, 1978) and V. G. Kiernan, *State and Society in Europe, 1550-1650* (Oxford: Blackwell, 1980). For a more recent examination in English, see P. H. Wilson, *Europe's Tragedy: A History of the Thirty Years War* (London: Penguin, 2010).

15 H. Van der Wee, 'The Influence of Banking on the Rise of Capitalism in North-West Europe, Fourteenth to Nineteenth Century', in A. Teichova, G. Kurgan-van Hentenryk and D. Ziegler (eds.), *Banking, Trade and Industry: Europe, America and Asia from the Thirteenth to the Twentieth Century* (Cambridge: Cambridge University Press, 1997), p. 180.

The Scientific Revolution: Bohemia and the Netherlands

The Thirty Years War enhanced the historically momentous partition of Europe into the unevenly developing, geographically ill-designated and differently fated West and East. This division profoundly affected not only the political, social and economic spheres, but also the pursuit of natural knowledge.

Before 1618 Bohemia, located in middle Europe, was the richest Habsburg dominion. It housed some of the great European mining centres: Jáchymov (Joachimsthal) and Kutná Hora (Kuttenberg). It was not a coincidence that the authors of the two classical texts on mining and metallurgy (reprinted for nearly 150 years) were active in those towns. Georgius Agricola (Georg Bauer) (1494-1550), who authored *De re metallica* (1556), was a physician in Jáchymov; Lazarus Ercker (?-1593), the author of *Beschreibung Allerfürnemisten Mineralischen Ertzt/vnnd Berckwercksarten etc.* (1574), was Warden and eventually Master of the Kutná Hora mint. He ultimately occupied the post of Chief Mining Officer of the Kingdom of Bohemia (*Oberster Bergmeister*).

As previously mentioned, the growth of the large-scale manorial economy, encompassing brewing, sheep-raising and carp-farming in purpose-built ponds made the 'Bohemian lords rich'. It was this milieu that inspired Hagecius to acquaint himself with brewing practice, which he describes in the pioneering booklet *De cerevisia etc.* (1585).

Deep mining and the construction of fish ponds depended on adequate surveying. Horský interestingly points to the basically identical methodologies and instruments employed in mines and fishpond surveying, on the one hand, and astronomy, on the other.[16]

On the eve of the Prague Castle defenestration, economic and intellectual omens in Bohemia appeared to augur well for a body politic, and for pursuits of natural knowledge, moving in the Dutch direction.

The clue as to why this development never came about lies in the difference between what the uprising/rebellion in Bohemia and the revolt in the Netherlands were about. Ostensibly Bohemia and the Netherlands faced a common foe: that is, the Catholic Habsburgs aspiring to dynastic dominance in Europe, albeit in two branches, the Austrian and the Spanish. Other intertwined factors – politics, economics and social structures – played their part in generating distinct outcomes, results which also affected the

---

16 Horský in *Prag um 1600*, p. 70.

scientific sphere. The time lag in the development of the Scientific Revolution in Bohemia, as compared with the Netherlands, is unambiguous.

For one thing, there was a difference in the role and interests of the nobility and the burghers in the two countries. Whereas, as a class, the nobility in Bohemia was strong and the burghers weak, in the Netherlands the situation was reversed. In Bohemia nothing like a Dutch-style defeudalisation/bourgeoisification had taken shape, nor did a noble-burgher inter-class alliance, which had characterised the Dutch fight against Spain, materialise.

When the Bohemian nobles, the elite social strata, raised the banner of rebellion, they were primarily concerned with their own socio-political interests. The dispute was over the nature of the monarchy, and Bohemia 'became one of the focal points of the conflict where a program of freedom of religion and the maintenance of liberties of the estates represented a defence against centralization'.[17] This is not to say, as Victor Kiernan seems to suggest, that Bohemia was virtually turning itself into a republic, but that its leaders shrank from the final step. As he himself points out:

> To set up a republic was an enterprise which only the most advanced and most revolutionary nations had courage for; the Dutch only half succeeded, the English only momentarily. Its time would come in the end far away from Europe, beyond the Atlantic.[18]

Indeed it would, but not at the same time, and under very different socio-political economic and cultural circumstances. Contrast Cook's appraisal with Smolka's analysis of scientific evolution in Bohemia from the middle of the sixteenth to the middle of the eighteenth century, by which time Newton and Linnaeus had arrived there.[19] Smolka's analysis specifically draws attention to the long-term negative impact that putting down the 1618 uprising had on the evolution of the Scientific Revolution in the Czech Lands. The Revolution was stifled

> in a quite decisive manner by the social situation which developed in the Czech area: re-catholization and the near-absolute ideological hegemony of the Jesuit Order, intellectual oppression and lack of freedom which went hand in hand with complete isolation from modern European scientific trends. And it is in a way paradoxical that whereas the historical conditions for scientific

---

17 J. Petráň and L. Petráňová, 'The White Mountain as a Symbol in Modern Czech History', in Teich (ed.), *Bohemia*, p. 144.
18 Kiernan, *State and Society in Europe*, p. 195.
19 H. J. Cook, 'The New Philosophy in the Low Countries', in Porter and Teich (eds.), *The Scientific Revolution*, pp. 137-38.

development at the beginning of the period involved were relatively favourable, at its end, when the Enlightenment and rationalism were beginning in Europe, Bohemia and Moravia were as it were starting from scratch, from the very beginning, but only slowly, hesitatingly and uneasily ... [20]

As it happened, there was a Dutch dimension to this eventual reform in the person of Gerard van Swieten, Empress Maria Theresa's physician. On his advice, medical instruction was restructured along the lines developed in Leiden by his influential teacher Herman Boerhaave (1668-1738). Moreover, and no less crucially, it was due to van Swieten that the Jesuit hold on education and censorship was broken.

Fig. 15 Emperor Franz Stephan (sitting) together with his natural science advisors. From left to right: Gerard van Swieten, Johann Ritter von Baillou (naturalist), Valentin Jamerai Duval (numismatist) and Abbé Johann Marcy (Director of the Physical Mathematical Cabinet).

Context and content of science

In a notable collection investigating the historical relationship between *Instruments, Travel and Science*, the editors go out of their way to point out that their historiographical approach is 'contextual'. Over the last two decades, they state

historians of science engaged in writing contextually have argued against a view of the development of knowledge, and particularly of scientific knowledge, as a unilinear process, and against the notion that the universality of science progressively imposes itself by the sheer force of the uniformity of the laws

---

20 Smolka, 'Bohemia', in Porter and Teich (eds.), *The Scientific Revolution*, pp. 234-35.

of nature. Conversely, these historians have described through multiple case studies how the universal dimension of science, was, in fact, predicated upon the context of its making and deeply grounded in locality.[21]

These observations should not obscure the fact that the contextual approach to the history of science has a past stretching beyond 1980. To all intents and purposes, it goes back to Hessen's seminal analysis (1931) of *Principia* against the background of the period in which Newton worked and lived. To ignore Needham's *Science and Civilisation in China* (1954-) as a work of contextual history is to fly in the face of the evidence. Written under Marxist influence, these and such-like studies were suspect *ab ovo*. Even scholars who acknowledge the relevance of the social context for the historiography of science feel that novel forays, such as Shapin and Schaffer's *Leviathan and the Air Pump*, have not delivered. That is, the authors 'place the history of science within a social context, but they did not discuss the possible influence of the content of science'.[22] As it happens, Schaffer's contribution to the volume on *Instruments, Travel and Science* illustrates that the relationship between the social context and the content of science is not reducible to a point-to-point relationship, but is mediated. Understanding it requires a more nuanced discussion based, for example, on the interdisciplinary approach exemplified below.

The subject matter of Schaffer's piece is the English gold trade in the seventeenth and early eighteenth centuries and the metrological concerns of Robert Boyle and Isaac Newton, protagonists of the gold system. This system crucially depended on gold, as a commodity, to be 'true gold, perfect metal' (in the words of John Locke).[23] Gold must have first attracted the attention of humans in prehistoric times. The use of sheep skins in the washing of gold-bearing sands to return gold particles may have given rise to the myth of the Golden Fleece. The notion of gold's superiority to other metals, be it in economy or medicine, is alchemical.

It is well established that Boyle's, Locke's and Newton's alchemical interests did not prevent them from looking for and obtaining 'hard' knowledge of nature. What Schaffer shows is that state-sponsored and commercial natural

---

21 M.-N. Bourget, Ch. Licoppe and H. O. Sibum (eds.), *Instruments, Travel and Science: Itineraries of Precision from the Seventeenth to the Twentieth Century* (London and New York: Routledge, 2002), p. 3.

22 M. J. Osler, 'The Canonical Imperative: Rethinking the Scientific Revolution', in M. J. Osler (ed.), *Rethinking the Scientific Revolution* (Cambridge: Cambridge University Press, 2000), p. 19.

23 S. Schaffer, 'Golden Means: Assay Instruments and the Geography of Precision in the Guinea Trade', in Bourget et al. (eds.), *Instruments*, p. 22.

history mattered to them too. That is, the development of precision methods to distinguish between counterfeits and true gold, shipped from the Gold Coast of Guinea to the Royal Mint in London where Warden Newton ruled supremely:

> Newton constructed a ferocious regime of governance within the institutions of natural philosophy and the walls of the Tower. In Mint work he insisted on accurate weighings and corrected what he judged an unacceptably large tolerance of error in the average weight of coins, called 'the remedy'. The Newtonian mint became an emblematic site of administrative metrology.[24]

Ever since Hessen's paper was published, it has been criticised, by non-Marxist historians as well as a number of Marxist ones, for its socio-economic 'determinism', 'reductionism' and suchlike. This points to an appreciable misreading of Hessen's aim, which was 'to determine the *basic tendency* [MT] of the interests of physics during the period immediately preceding Newton and contemporary with him'.[25]

Turning to Schaffer's article, one has the impression that he owes a good deal to Hessen, if not to Marx. Newtonian projects, he states,

> were not, of course, limited to fiscal standardisation. Between 1709 and 1713, ably assisted by the young Cambridge mathematician Roger Cotes, Newton prepared a revised edition of his *Principia mathematica*. Some 'hypotheses' which had prefaced its third book in 1687 now became reworked as *'regulae philosophandi'*. The second rule stated that 'the causes assigned to natural effects of the same kind must be, so far as possible, the same. Examples are … the falling of stones in Europe or America'. The *Principia*, in this sense, was a handbook for travellers. Newton described the celebrated marvels of tidal ebb and flow in the East Indies, the Straits of Magellan and the Pacific. Keen to show the universal grip of his gravitational model of lunar pull, Newton here faced characteristic troubles of trust in travellers' tales. Against Leibnizian rivals, Newton and Cotes now sought massively to reinforce the apparent precision of their measures. They discussed whether to omit or include tide data from variably reliable mariners using assumptions about such parameters as the earth's destiny. The link between trust in persons and in creation's constancy was even clearer in their work on the length of isochronic pendulums in Europe, American and in Africa too. In the 1680s Newton had hoped that 'the excess of gravity in these Northern places over the gravity at the Equator' would be 'determined exactly by experiments conducted with greater diligence'. Cotes now 'considered to make the Scholium appear to the best advantage as to

---

24 Ibid., p. 39.
25 B. Hessen, 'The Social and Economic Roots of Newton's "Principia"', in *Science at the Cross Roads*, 2nd ed. (London: Cass, 1971), p. 165.

the numbers'. They would make a table of the variations in the length of a seconds pendulum at different points on earth, visibly accurate over very small length differences of fractions of an inch. Cotes held 'that the generality of Your Readers must be gratified with such trifles, upon which the commonly lay ye greatest stress'.[26]

Interdisciplinarity in Becher's thought

Boyle, Locke and Newton shared their fascination with the historically linked fields of mining, metallurgy, alchemy, chemistry and coinage with continental contemporaries such as Johann Joachim Becher (1635-?).[27] In general histories dealing with the rise of modern chemistry, he usually has a place in the discussion of the origins of the doctrine of phlogiston. In works surveying the evolution of economic thought, Becher is mentioned as the leading representative of German (Austrian) mercantilism.

Among the most striking features of Becher's life is his continuous fascination with alchemy. Becher emerges as someone who unquestionably believed that transmutation was possible though he had little regard for the pursuit of an elixir of life. However, scepticism regarding this or that aspect of alchemy did not hinder Becher from re-iterating that without alchemy, metallurgy cannot be fully understood. Indeed, he goes on, anyone who is opposed to alchemy should realise that he either does not comprehend or does not fully appreciate that metallurgy and coinage form the foremost part of his prince's income.

Becher's reference to the sphere of coinage touches on an element of German political and economic history which should not be lost sight of. That is, the non-uniformity of the German system of coinage, being partly the product of the political and economic division of Germany but also contributing to it, long before and after the Thirty Years War.

One of the problems arising out of these coinage conditions in Germany was the exchange of currencies between the independent territories within the framework of the Holy Roman Empire of the German Nation, as well as between these territories and the non-German states. Discussing

---

26 Schaffer, 'Golden Means', pp. 37-8.
27 See G. Frühsorge and G. F. Strasser (eds.), *Johann Joachim Becher (1635-1682)* (Wiesbaden: Harrassowitz, 1993). This volume includes my 'Interdisciplinarity in J. J. Becher's Thought' on which I draw here (pp. 23-40). It was previously published in *History of European Ideas*, 9 (1988), 145-60.

money-exchange in the section on coinage in *Politische Discurs*, Becher proposes to control it institutionally by establishing a discount-house (*Wexel-Banck*).[28]

Underlying this proposal was Becher's preoccupation with the movement of money and commodities and their relation to each other. The interest of this lies in the fact that Becher ascribes to money the function of a commodity. Becher may be among the first writers to espouse such an idea. Here is what he says: '... thus in respect of a country there is no commodity as expensive and necessary as money; and no commodity going out of the country should be charged with more duty than money because money is equally the nerve and the soul of the country'.[29]

This proposition ties in with the impact of merchant capital felt at the centre of Europe as part of the transformation from feudal to capitalist economy, made while Becher was composing his substantive work. An inherent feature of this process was the growing production of commodities to be sold, that is exchanged for money, on the market at home and abroad. Becher was concerned precisely with such questions in the context of Germany's political and economic ruin after the Thirty Years War. That is, he was galvanised by with the ascendancy of merchant capital in the economic life of this geographical area, a process characterised by the interlocking circulation of commodities and money. There is first-hand evidence for Becher spelling out the idea of the circuit of merchant capital. It is contained in his appraisal of merchant-manufacturers as the mainstay of the community:

> Now finally regarding the point about *consumption*, I wish to add in praise and honour of the *Verlaeger* [merchant-manufacturers] the following. Namely that they are solely to be regarded as the pillars of the three estates because the artisan lives from them, from him the nobleman, from him the Prince, and from all these again the merchant. These are the hands which have to join.[30]

---

28 J. J. Becher, *Politische Discurs von den eigentlichen Ursachen des Auff- und Abnehmens der Staedt, Laender und Republicken* (Frankfurt am Main: Zunner, 1688), p. 269. This is the third edition of Becher's arguably most renowned publication, apart from *Physica Subterranea*. The first edition, *Politischer Discours*, appeared in 1668 and the second edition in 1673. The original title of *Physica Subterranea* was *Actorum Laboratorii Chymici Monacensis, Seu Physicae Subterraneae Libri Duo* (Frankfurt am Main: J. D. Zunneri, 1669). Despite the title it was a one-volume book. A second edition with three previously published supplements (1671, 1675, 1680) appeared in 1681. It was Becher himself who translated his own opus into German and published it as *Chymisches Laboratorium Oder Unter-erdische Naturkuendigung* (Frankfurt am Main: J. Haass, 1680), along with the first and second supplements and another publication *Ein Chymischer Raetseldeuter*. A second edition appeared in 1690.

29 Becher, *Politische Discurs*, p. 269.

30 Ibid., p. 106.

Becher's thought reveals 'cycle-mindedness' as permeating his natural philosophy just as much as his economic thinking. For Becher, unaffected by growing specialisation, it became the concept through which the natural and economic worlds interrelated. Cycle-mindedness and circulation-mindedness were terms employed by Joseph Needham when he called attention to the Taoist appreciation of cyclical change. At the same time he highlighted the eminent role of the notions of 'circle' and 'circulation' in the intellectual life of the Renaissance, when the foundations of what is called modern science were laid. The scholar to whose stimulating work we owe the most insight is Walter Pagel, but in these matters he appears not to have become as influential as he did in others.[31] Cycle-mindedness is bound up with the ideas of 'recurrence' and 'continuity'. In fact, it constitutes one of the most enduring visions of being from prehistoric times to the present, emerging in the belief in life after death or in the pursuit of perpetual motion machines. Such things, we know, were of concern to Becher. They should be viewed in the context of his belief in circulation – grounded in his reading and literal interpretation of the Bible – as the underlying principle of the natural and economic order of things.

But beyond that, Becher probably owes the idea of circulation to his alchemical and chemical interests. Can one doubt that he was familiar with the alchemical symbol of the serpent with his tail in his mouth? The circular form of the *Uroboros* was taken by the alchemists to stand for the 'death' and 'resurrection' of matter undergoing eternal chemical change. Moreover, since the sixteenth century at least, the terms 'distillation' and 'circulation' have come to be used analogically. There is direct evidence, in fact, for Becher making use of this analogy when he stresses that nature is in a state of perpetual motion, that this motion is circular and that it can therefore be compared to distillation.[32]

---

31 J. Needham, *The Grand Titration: Science and Society in East and West* (London: Allen & Unwin, 1969), pp. 227-28; W. Pagel, *William Harvey's Biological Ideas: Selected Aspects and Historical Background* (Basel and New York: Karger, 1967).

32 Becher, 'Revera autem, perpetua haec circulatio destillationi Chymicae comparari potest', *Physica Subterranea*, p. 50; idem, 'Nun kam aber? in der That dieser staete Cirkelgang mit einer Chymischen *distillation* verglichen werden', *Chymisches Laboratorium*, p. 131.

Epilogue

Becher was a polymath whose interests included human and veterinary medicine, mathematics, physics, chemistry, education, philology, technology, husbandry, political economy, social organisation, colonialism and natural philosophy. It is the sheer breadth of his inquiries that demands an interdisciplinary approach to the assessment of Becher's place in the world of seventeenth-century learning.

As illustrated above, Becher had a pretty good understanding of the dominating role of mercantile activity in the society in which he lived and was involved. As an economist and chemist, Becher was superior to Leibniz, with whom he is occasionally juxtaposed;[1] that said, Becher certainly was inferior in mathematics and physics to his great polymath contemporary, the fields in which Leibniz was originally and primarily active. Where a useful comparison may be attempted is in the ways Becher and Leibniz read the Bible and thought about God in the context of investigating natural phenomena.[2]

Stripped down to its essentials, Becher's philosophy of nature may be summarised in the following way. God is the first Being, a Being who is not created but who creates everything. As such, He has created nature as the vehicle of orderly motion. The primary stuff or raw material of nature is earth, of which everything is ultimately composed and into which everything is eventually decomposed. In Becher's opinion, God is a chemist who made

---

1 H. Breger, 'Becher, Leibniz und die Rationalität', in G. Frühsorge and G. F. Strasser (eds.), *Johann Joachim Becher (1635-1682)* (Wiesbaden: Harrassowitz, 1993), pp. 69-84. See also Pamela Smith's valuable juxtaposition of Becher and Leibniz in her *The Business of Alchemy: Science and Culture in the Holy Roman Empire* (Princeton, NJ: Princeton University Press, 1994).

2 H. Rudolph, 'Kirchengeschichtliche Beobachtungen zu J. J. Becher', *Becher*, pp. 173-96; M. Stewart, *The Courtier and the Heretic: Leibniz, Spinoza, and the Fate of God in the Modern World* (New Haven, CT and London: Yale University Press, 2005).

http://dx.doi.org/10.11647/OBP.0054.07

nature, from the creation of the world to its end, a cycle of transformation of one kind of earth into another.

In Leibniz's speculation, all-provident God ordained that the world He brought into being was to exist in harmony. Once created, it was to be the best of possible worlds, not needing His continuous intervention to keep it running like clockwork. This harmony was ultimately due to multiple primary entities ('monads') endowed with 'soul' that combined to form all matter.

Au fond, Becher and Leibniz were confronted with squaring the circle, as were Boyle, Descartes, Newton, Linnaeus and scores of other seventeenth- and eighteenth-century explorers of nature. They had to harmonise their findings and explanations of natural phenomena with the words of Scripture. To ignore this maxim was perilous in the light of Giordano Bruno's burning at the stake (1600) and Galileo's trial (1633). When Descartes learned of Galileo's condemnation by the Inquisition, he suppressed his treatise on *The World* (1634).

Protestant Christianity could be no more tolerant than the Roman Holy Office. In effect, the Reformation had reinforced the Bible as the undisputable authority. The Spanish physician Michael Servetus, whose inklings of the lesser circulation of blood were contained in his unorthodox *Restitutio Christianismi* (denial of Christ's divinity), was burnt in Geneva at Calvin's behest (1553). Newton was very careful when it came to concealing his heretical Arianism. And the deeply religious Linnaeus, who believed in God, trusted in the Bible as the Word of God and saw himself as God's interpreter of nature, was, for all that, not an orthodox Christian. So much so that, after Linnaeus's death, his son had to defend him in the face of accusations of atheism.[3]

The burning of suspected or real heretics was the uncompromising reaction of a Christianity that had become the dominant intellectual and cultural force in Europe since it was instituted as the state religion of the *imperium Romanum* in the fourth century. Paradoxically, as the eminent historian of science Richard Westfall put it, the actual result of the many treatises demonstrating the existence of God from natural phenomena has been the separation of science from revealed theology. Westfall compared

---

3   For religious views of Linnaeus, see contributions by Sten Lindroth and Tore Frängsmyr in T. Frängsmyr (ed.), *Linnaeus: The Man and his Work* (Berkeley, Los Angeles, CA and London: University of California Press, 1983). K. Hagberg's *Carl Linnaeus*, transl. Alan Blair (London: Cape, 1952) still provides useful information on the religious as well as economic and political dimensions of the Swedish naturalist's activities. For a more recent treatment, see L. Koerner, *Linnaeus: Nature and Nation* (Cambridge, MA: Harvard University Press, 1999).

two incidents supporting his point about the assertion of science's autonomy by the end of the seventeenth century:

> Early in the seventeenth century, the Catholic Church, under the leadership in this respect of Cardinal Bellarmino, condemned Copernican astronomy because it conflicted with the overt meaning of certain passages of Scripture. Sixty-five years later Newton engaged in a correspondence with Thomas Burnet about Burnet's *Sacred Theory of the Earth*. Burnet had convinced himself that the Scriptural account of the creation was a fiction, composed by Moses for political purposes, which could not possibly be true in a philosophical sense. In the correspondence, Newton defended the truth of Genesis, arguing that it stated what science (chemistry in this case) would lead us to expect. Where Bellarmino had employed Scripture to judge a scientific opinion, both Burnet and Newton used science to judge the validity of Scripture. To speak merely of the autonomy of science does not seem enough; we need to speak rather of its authority, to which theology had now become subordinate. The positions of the two had been reversed. That change also has never been reversed anew.[4]

The Christian Westfall saw in the new relation between science and Christianity 'one dimension of the new order that the Scientific Revolution ushered into being ... one more reason why I will not part with the concept'. He was convinced that the Scientific Revolution 'is the key not only to the history of science but to modern history as well'. Indeed, Westfall argued that the separation of the Western world from the rest of the globe stemmed from the growing scientific foundation that European technology had acquired since the seventeenth century.

Westfall appears to have believed that no one had seriously addressed the question 'of what it was about the new science that made it adaptable to technological use'. He identified three distinguishing features: its quantitative character, the employment of experimentation and the development of a more satisfactory account of nature.

In more general terms, the question was in fact addressed by Marx, 'that most astute historian of nineteenth-century technology (and Manchester)'.[5] He did so in a preparatory work (1857-1858), published for the first time under the title *Grundrisse der Kritik der politischen Ökonomie (Rohentwurf)*

---

4 R. S. Westfall, 'The Scientific Revolution Reasserted', in M. J. Osler (ed.), *Rethinking the Scientific Revolution* (Cambridge: Cambridge University Press, 2000), p. 50. For the following quotations, see pp. 50-1.

5 This observation was made by a much respected non-Marxist historian of science. See J. V. Pickstone's thoughtful synthesis of the state-of-the-art scholarship in the history of science, technology and medicine, *Ways of Knowing: A New History of Science, Technology and Medicine* (Manchester: Manchester University Press, 2000), p. 24.

in 1939-1941. As is known or should be known, the chief aim of Marx's theoretical endeavour was to discover the laws underlying the formation and development of capital and thus to provide the key to comprehending capitalism as a historically evolved system of social production. It is in this connection that Marx encountered the problem of the role of science and technology in the development of the productive forces under capitalism.[6]

In the *Grundrisse* Marx touches on the problem, among other matters, when he analyses categories such as labour process, fixed capital, machine, etc. Marx notes that

> Nature builds no machines, no locomotives, railways, electric telegraphs, self-acting mules, etc. These are products of man-made industry ... they are *organs of the human brain, created by the human hand*, a power of knowledge objectified. The development of fixed capital [appearing as a machine] reveals to what degree general social knowledge has become a *direct force of production*.

On following Marx's later thinking on this subject in *Capital* (1867), his major published work, we find that by then he regarded science as an intellectual power (*geistige Potenz*) of the process of production rather than as a direct force of production. Marx frequently refers in *Capital* to *Produktivkraft der Arbeit* in the sense of productivity and writes that it is 'determined by various circumstances, among others, by the average skill of the workmen, the state of science and the degree of its practical application, the social organization of production, the extent and capabilities of the means of production and by physical conditions'.

No doubt what happened was that Marx, on mature reflection, concluded that a more subtle relationship between science and production existed than the point-to-point one he had previously established. It has been my view for a long time that practitioners as well as theoreticians concerned with this issue could benefit from Marx's later approach. That is, to recognise that the relationship between science and production is a mediated one, depending on factors such as military needs, economic expectations, technological feasibility, political interests and others. Looked at it in this way, scientific knowledge represents a *potential* rather than a direct force of production.

This assigns to mediation a centrally important epistemological function; it helps us understand the part played by the interaction of a plurality of

---

6 What follows draws on my 'The Scientific-Technical Revolution: An Historical Event in the Twentieth Century?', in R. Porter and M. Teich (eds.), *Revolution in History* (Cambridge: Cambridge University Press, 1986), pp. 317-30.

factors in transitional phases of history. It is in this context that it pays to return to Westfall's reaffirmation of the concept of the Scientific Revolution, one with which I agree in principle but do not find adequate.

Take Westfall's observation that the Scientific Revolution brought into being a new order and, moreover, that it provides a guide not only to the history of science but to modern history. For one thing, it is not clear what he means by 'new order' – is it political, social, economic, intellectual, conceptual, methodological, epistemic, etc.? True, Westfall points to the subordination of theology to science as one of its hallmarks, in addition to the closer relation of science and technology. But his lack of interest in placing the Scientific Revolution within the wider, plural history of the period he discusses (1543-1687) is evident.

Westfall's comparative indifference contrasts with the view of the editor of a volume on Galileo, who emphasises that the development of science is affected by external conditions in which it is pursued:

> What cannot be in doubt is that between, say somewhat arbitrarily, the dates of 1543 and 1687, many things had radically changed and the world was, and was further becoming, a widely different kind of place. Science, as any other human endeavour, does not exist in a vacuum. It is not an isolated, independent system of thought and practice. What happens in other realms affects how science is practiced, perceived, and received.[7]

Alas, this perspective is not reflected in the actual contributions to the valuable, if traditional, collection – apart from the attention it pays to Church and Scripture.

As stated at the beginning of this book, my theme is the Scientific Revolution as a distinctive movement of thought and action that came into its own in certain European countries by the seventeenth century. At that point, the Greek 'inquiry concerning nature' had acquired the form that has since been universally adopted.

Some ways of knowing, like observation and experience, go back to the times when ape-like creatures began to employ them purposefully and, in the process, became recognisably human. Others – classification, systematisation and theorising – were developed in classical antiquity. Systematic experimentation and quantification represent procedures of

---

7 'Introduction', in P. Machamer (ed.), *The Cambridge Companion to Galileo* (Cambridge: Cambridge University Press, 1998), p. 3.

investigating and comprehending nature that began to materialise in Europe during the late Middle Ages and the Renaissance.

By and large, historians have come to identify this phase as the transition from the late medieval to the early modern period in Europe. Some, influenced by Marxism, view it as the opening stage of transition from feudalism to capitalism in Europe, with merchant capital gaining control of commerce and manufacture.

It is in this intellectual context that John Desmond Bernal wrote of the rise of capitalism and the birth of modern science, in about the middle of the fifteenth century, as related processes, arguing in his seminal *The Social Function of Science* (1939):

> Though capitalism was essential to the early development of science, giving it, for the first time, a practical value, the human importance of science transcends in every way that of capitalism, and, indeed, the full development of science is incompatible with the continuance of capitalism.[8]

The reading of this work in 1940 was pivotal in arousing my interest in the history and philosophy of science in general, and in the Marxist materialist conception of history in particular. It led to my earliest publication – on the eve of my taking the first degree – in *Nature* (1944).[9] Some seven decades later, I still find this approach to history enlightening although not dogmatically prescriptive.

---

8 J. D. Bernal, *The Social Function of Science*, 3rd ed. (London: Routledge, 1942), pp. 408-09. See my 'J. D. Bernal: The Historian and the Scientific Technical Revolution', *Interdisciplinary Science Reviews*, 33 (2008), 135-39.

9 N. Teich, 'Influence of Newton's Work on Scientific Thought', *Nature*, 153 (1944), 42-5.

References

Abulafia, D., 'The Impact of Italian Banking in the Late Middle Ages and the Renaissance, 1300-1500', in Teichova, A., Kurgan-van Hentenryk, G., and Ziegler, D. (eds.), *Banking, Trade and Industry: Europe, America and Asia from the Thirteenth to the Twentieth Century* (Cambridge: Cambridge University Press, 1997), pp. 17-34.

Anderson, P., *Lineages of the Absolute State* (London: New Left Books, 1974).

Andrewes, W. J. H., 'Time and Clocks', in Maran, Stephen P. (ed.), *The Astronomy and Astrophysics Encyclopaedia* (New York: John Wiley & Sons, 1991) pp. 929-31.

Anscombe, E. and Geach, P. T. (eds. and transl.), *Descartes: Philosophical Writings*, with an Introduction by A. Koyré (Sunbury-on-Thames: Nelson, 1976),

Aristotle, *Politics*, I, ii, 9-11 (Loeb Classical Library, Vol. 21, transl. H. Rackham) (Cambridge, MA: Harvard University Press and London: Heinemann, 1977).

Bacon, F., 'Aphorisms – Book One', in *The New Organon and Related Writings*, ed. by F. H. Anderson (New York: Macmillan/Library of Liberal Arts, 1987).

Bauer, L. and Matis, H., *Geburt der Neuzeit. Vom Feudalsystem zur Marktgesellschaft* (Munich: Deutscher Taschenbuch-Verlag, 1988).

Bayertz, K., 'Über Begriff und Problem der wissenschaftlichen Revolution', in Bayertz, K. (ed.), *Wissenschaftsgeschichte und wissenschaftliche Revolution* (Hürth-Efferen: Pahl-Rugenstein, 1981), pp. 11-28.

— 'Was ist moderner Materialismus', in Bayertz, K., Gerhard, Myriam and Jaeschke, W. (eds.), *Weltanschauung, Philosophie und Naturwissenschaft im 19. Jahrhundert*, Vol. 1: *Der Materialismus-Streit* (Hamburg: Meiner, 2007), pp. 50-70.

Becher, J. J., *Actorum Laboratorii Chymici Monacensis, Seu Physicae Subterraneae Libri Duo* (Frankfurt am Main: J. D. Zunneri, 1669).

— *Chymisches Laboratorium Oder Unter-erdische Naturkuendigung* (Frankfurt am Main: J. Haass, 1680), with first and second supplements and another publication, *Ein Chymischer Raetseldeuter*. A second issue appeared in 1690.

— *Politische Discurs von den eigentlichen Ursachen des Auff- und Abnehmens der Staedt, Laender und Republicken* (Frankfurt am Main: Zunner, 1688).

Bedini, S. A. and Solla Price, D. J. da, 'Instrumentation', in Kranzberg, M. and Pursell, C. W., Jr. (eds.), *Technology in Western Civilization*, Vol. 1 (New York: Oxford University Press, 1967), pp. 168-87.

Beretta, M., Clericuzio, A. and Principe, L. M. (eds.), *The Accademia del Cimento and its European Context* (Sagamore Beach, MA: Science History Publications, 2009).

Berg, M., *The Age of Manufactures 1700-1820: Industry, Innovation and Work in Britain*, 2nd ed. (London and New York: Routledge, 1994), http://dx.doi.org/10.4324/9780203990971

Bernal, J. D., *The Social Function of Science*, 3rd ed. (London: Routledge, 1942).

— *The Extension of Man: A History of Physics Before 1900* (London: Weidenfeld and Nicholson, 1972).

Birch, T., *The History of the Royal Society of London for Improving of Natural Knowledge from Its First Rise, in which the Most Considerable of Those Papers Communicated to the Society, which Have Hitherto Not Been Published, are Inserted as a Supplement to the Philosophical Transactions*, 4 vols. (London: printed for A. Millar, 1756-1757).

Boas Hall, M., 'Science in the Early Royal Society', in Crosland, M. (ed.), *The Emergence of Science in Western Europe* (London and Basingstoke: Macmillan, 1975), pp. 57-77.

Bohadsch, J. T., *Beschreibung einigen in der Haushaltung und Färbekunst nutzbaren Kräutern, die er in seinen durch drey Jahre unternommenen Reisen im Königreich Böheim entdecket hat* (Prague: Franz Ignatz Kirchner, 1755).

— *Abhandlung vom Gebrauch des Waides in der Haushaltung* (Prague: [n. pub.], [n.d.]).

— *Dienst- und Nutzbarer Patriotischer Vorschlag, wienach dem Königreich Böheim ein ungemeiner Vortheil von sonderbarer Beträchtlichkeit jährlich zuwachsen könnte* (Prague, 1758).

Born, I. v., *Schreiben des Herrn Ignatz von Born … an Herrn Franz Grafen von Kinsky, Ueber einen ausgebrannten Vulkan bey der Stadt Eger in Boehmen* (Prague: Gerle, 1773).

— 'Antwort des Herrn von Born, auf das Schreiben des Herrn Grafen von K....', *Abhandlungen*, 1 (1775), 253-63.

Bourget, M.-N., Licoppe, Ch. and Sibum, H. O. (eds.), *Instruments, Travel and Science: Itineraries of Precision from the Seventeenth to the Twentieth Century* (London and New York: Routledge, 2002), http://dx.doi.org/10.4324/9780203219010

Bowler, P. J. and Rhys Morus, I., *Making Modern Science: A Historical Survey* (Chicago, IL and London: University of Chicago Press, 2005).

Breger, H., 'Becher, Leibniz und die Rationalität', in Frühsorge, G. and Strasser, G. F. (eds.), *Johann Joachim Becher (1635-1682)* (Wiesbaden: Harrassowitz, 1993), pp. 69-84.

Brooke, J. H., *Science and Religion: Some Historical Perspectives* (Cambridge: Cambridge University Press, 1991).

Bryson, B. (ed. and intr.), *Seeing Further the Story of Science and the Royal Society* (London: HarperPress, 2010).

Buridan, J., *The Impetus Theory of Projectile Motion*. Transl., intr., and annotated by Marshall Clagett, in Grant, E. (ed.), *A Source Book in Medieval Science* (Cambridge, MA: Harvard University Press, 1974), pp. 275-80.

Bylebyl, J. J., 'Nutrition, Quantification and Circulation', *Bulletin of the History of Medicine*, 51 (1977), 369-85.

Bynum, W. F., Brown, E. J. and Porter, R. (eds.), *Dictionary of the History of Science* (London and Basingstoke: Macmillan, 1981).

Cardwell, D., *The Fontana History of Technology* (London: Fontana Press, 1994).

Chilvers, C. A. J., 'The Dilemmas of Seditious Men: The Crowther-Hessen Correspondence in the 1930s', *The British Journal for the History of Science*, 36 (2003), 417-35, http://dx.doi.org/10.1017/S0007087403005156

— 'La signification historique de Boris Hessen', in Gerout, S. (ed.), *Les Racines sociales et économiques des Principia de Newton* (Paris: Vuibert, 2006), pp. 179-206.

Cipolla, C. M., *The Economic History of World Population* (Harmondsworth: Penguin, 1962).

— 'Introduction', in Cipolla, C. M. (ed.), *The Fontana Economic History of Europe: The Industrial Revolution* (London and Glasgow: Collins/Fontana Books, 1973).

Clagett, M. (ed.), *Critical Problems in the History of Science* (Madison, WI: University of Wisconsin Press, 1959).

Cohen, H. F., *The Scientific Revolution: A Historiographical Inquiry* (Chicago, IL and London: University of Chicago Press, 1994).

Cohen, I. B., 'The Eighteenth-Century Origins of the Concept of Scientific Revolution', *Journal of the History of Ideas*, 37 (1976), 257-88.

— *The Revolution in Science* (Cambridge, MA: Belknap Press, 1985).

Coleman, D. C., *Myth, History and the Industrial Revolution* (London and Rio Grande, OH: Hambledon Press, 1992).

Comenii, J. A., *Via Lucis J. A. Komenského Cesta světla* (Prague: Státní Pedagogické Nakladatelství, 1961).

Comenius, J. A., *Via lucis, vestigata et vestiganda, hoc est rationabilis disquisitio, quibus modis intellectualis animorum Lux, Sapientia, per omnes omnium hominum mentes, et gentes, jam tandem sub mundi vesperam feliciter spargi posit. Libellus ante annos viginti sex in Anglia scriptus, nunc demum typis exscriptus et in Angliam remissus* (Amsterdam: Apud Christophorum Cunradu, 1668).

— *The Way of Light*, trans. E. T. Campagnac (Liverpool: Liverpool University Press, 1938).

Cook, H. J., 'The New Philosophy in the Low Countries', in Porter, R. and Teich, M. (eds.), *The Scientific Revolution in National Context* (Cambridge: Cambridge University Press, 1991), pp. 115-49, http://dx.doi.org/10.1017/CBO9781139170215.005

— *Matters of Exchange: Commerce, Medicine, and Science in the Dutch Golden Age* (New Haven, CT and London: Yale University Press, 2007), http://dx.doi.org/10.12987/yale/9780300117967.001.0001

Corfield, P. J., *Time and the Shape of History* (New Haven, CT and London: Yale University Press, 2007), http://dx.doi.org/10.12987/yale/9780300115581.001.0001

Crombie, A. C., *Augustine to Galileo: The History of Science, A.D. 400-1650* (London: Falcon Press, 1952).

— *Robert Grosseteste and the Origins of Experimental Science, 1100-1700* (Oxford: Clarendon Press, 1953).

— 'Contribution to Discussion of Part Three: Science and Technology in the Middle Ages', pp. 272-91, in Crombie, A. C. (ed.), *Scientific Change, Symposium on the History of Science, University of Oxford 9-15 July 1961* (London: Heinemann, 1963), pp. 316-23.

— 'Historians and the Scientific Revolution', *Physis: Rivista Internazionale di Storia della Scienza*, 11 (1969), 167-80.

— *Styles of Scientific Thinking in the European Tradition: The History of Argument and Explanation Especially in the Mathematical and Biomedical Sciences and Arts* (London: Duckworth, 1994), Vol. 1.

Crosby, A. W., *The Measure of Reality: Quantification and Western Society, 1250-1600* (Cambridge: Cambridge University Press, 1998).

Crowther, J. G., *The Social Relations of Science* (New York: The Macmillan Company, 1942).

Cunningham, A., 'William Harvey and the Discovery of the Circulation of the Blood', in Porter, R. (ed. and intr.), *Man Masters Nature: 25 Centuries of Science* (London: BBC Books, 1987), pp. 65-76

Cunningham, A. and Williams, P., 'De-centring the "Big Picture": The Origins of Modern Science and The Modern Origins of Science', in *The British Journal for the History of Science*, 26/4 (December 1993), pp. 407-32, http://dx.doi.org/10.1017/s0007087400031447

Dainton, B., 'Past, What Past?', *The Times Literary Supplement*, 8 January 2010, 9.

Damerow, P., Renn, J., Rieger, S. and Weinig, P., *Mechanical Knowledge and Pompeian Balances*, Preprint 145 (Berlin: Max-Planck-Institut für Wissenschaftsgeschichte, 2000).

Day, J., 'Shorter Notices', *The English Historical Review*, Vol. 109/432 (1994), 701, http://dx.doi.org/10.1093/ehr/cix.432.701-a

Dear, J. P., *Revolutionizing the Sciences: European Knowledge and its Ambitions, 1520-1700* (Basingstoke: Palgrave, 2001).

Descartes, *Descartes: Philosophical Letters*, ed. and transl. by Kenny, A. (Oxford: Clarendon Press, 1970).

— 'Discourse on the Method', in Anscombe, E. and Geach, P. T. (ed. and transl.), *Descartes: Philosophical Writings*, with an Introduction by A. Koyré (Sunbury-on-Thames: Nelson, 1976), pp. 5-57.

— 'Description of the Human Body', in Gaukroger, S. (ed.), *Descartes: The World and Other Writings* (Cambridge: Cambridge University Press, 1998), pp. 170-205, http://dx.doi.org/10.1017/cbo9780511605727.009

Dijksterhuis, F. J., 'Constructive Thinking: A Case for Dioptrics', in Roberts, L., Shaffer, S. and Dear, P. (eds.), *The Mindful Hand: Inquiry and Invention from the Late Renaissance to Early Industrialization* (Amsterdam: Koninkliijke Nederlandse Akademie van Wetenschappen, 2007), pp. 59-82.

Drábek, P. (ed.), *Tadeáš Hájek z Hájku* (Prague: Společnost pro dějiny věd a techniky, 2000).

Dreyer, E. L., *Zheng He: China and the Oceans in the Early Ming Dynasty* (New York and London: Pearson Longman, 2007).

Efron, N. J., 'Liberal Arts, Eirenism and Jews in Rudolfine Prague', *Acta historiae rerum naturalium necnon technicarum. Prague Studies in the History of Science and Technology*, Vol. 1 (1997), 24-35.

Engels, F., 'Ludwig Feuerbach and the End of Classical German Philosophy', in Marx, K. and Engels, F., *Selected Works in Three Volumes*, Vol. 3 (Moscow: Progress Publishers, 1973), pp. 335-76.

Evans, J., *The History and Practice of Ancient Astronomy* (New York and Oxford: Oxford University Press, 1998).

Evans, R. J. W., *Rudolf II and His World: A Study in Intellectual History, 1576-1612* (Oxford: Clarendon Press, 1973).

— 'Rudolf II: Prag und Europa um 1600', in [no editor], *Prag um 1600: Kunst und Kultur am Hofe Kaiser Rudolf II*, Vol. 1 (Freren: Luca Verlag, 1988), pp. 27-37.

Fara, P., *Science: A Four Thousand Year History* (Oxford: Oxford University Press, 2009).

Farrington, B., *Francis Bacon: Philosopher of Industrial Science* (London: Macmillan; New York: Haskell House, 1973).

Field, J. V., 'Mathematics and the Craft of Painting: Piero della Francesca and Perspective', in Field, J. V. and James, Frank A. J. L. (eds. and intr.), *Renaissance and Revolution: Humanists, Scholars, Craftsmen and Natural Philosophers in Early Modern Europe* (Cambridge: Cambridge University Press, 1993), pp. 73-95.

Field, J. V. and James, Frank A. J. L. (eds. and intr.), *Renaissance and Revolution: Humanists, Scholars, Craftsmen and Natural Philosophers in Early Modern Europe* (Cambridge: Cambridge University Press, 1993).

Findlen, P., 'Cabinets, Collecting and Natural Philosophy', in Fučíková, E. et al., *Rudolf II and Prague* (Prague: Prague Castle Administration and London and Milan: Thames and Hudson, 1997), pp. 209-19.

Finlay, R., 'China, the West, and World History in Joseph Needham's Science and Civilisation in China', *Journal of World History*, 11 (2000), 265-303.

Finley, M. I., *The Ancient Greeks* (Harmondsworth: Penguin, 1977).

Folta, J. et al., *Dějiny exaktních věd v českých zemích do konce 19. století* [*History of the Exact Sciences in the Czech Lands up to the End of the Nineteenth Century*] (Prague: Nakladatelství Československé akademie věd, 1961).

Forbes, R. J., 'Metals and Early Science', *Centaurus*, 3 (1953-1954), 24-31, http://onlinelibrary.wiley.com/doi/10.1111/j.1600-0498.1953.tb00517.x/pdf

Frängsmyr, T. (ed.), *Linnaeus: The Man and his Work* (Berkeley, Los Angeles, and London: University of California Press, 1983).

— 'Linnaeus as a Geologist', pp. 110-55.

Frankenberger, Z., 'Jan Křtitel Boháč: život a dílo', *Věstník Královské české společnosti nauk*, 12 (1950), 1-122.

Freudenthal, G., 'The Hessen-Grossman Thesis: An Attempt at Rehabilitation', *Perspectives on Science*, 13/2 (Summer 2005), 166-93, http://dx.doi.org/10.1162/106361405774270575

Fritsche, J., 'The Biological Precedents for Medieval Impetus Theory and its Aristotelian Character', *The British Journal for the History of Science*, 44/1 (2011), 1-27. http://dx.doi.org/10.1017/S0007087410000774

Frühsorge, G. and Strasser, G. F. (eds.), *Johann Joachim Becher (1635-1682)* (Wiesbaden: Harrassowitz, 1993).

Gascoigne, J., 'The Royal Society, Natural History and the Peoples of the "New World(s)", 1660-1800', *The British Journal for the History of Science*, 42/155.4 (2009), 539-62.

Gaukroger, S., *Descartes: An Intellectual Biography* (Oxford: Oxford University Press, 1995).

Gaukroger, S. (ed.), *Descartes: The World and Other Writings* (Cambridge: Cambridge University Press, 1998), http://dx.doi.org/10.1017/cbo9780511605727

Gellner, E., 'Along the Historical Highway', *The Times Literary Supplement*, 16 March 1984, 279-80.

— 'Knowledge of Nature and Society', in Teich, M., Porter, R. and Gustafsson, B. (eds.), *Nature and Society in Historical Context* (Cambridge: Cambridge University Press, 1997), pp. 9-17.

Goodman, D., 'The Scientific Revolution in Spain and Portugal', in Porter, R. and Teich, M. (eds.), *The Scientific Revolution in National Context* (Cambridge: Cambridge University Press, 1992), pp. 158-77, http://dx.doi.org/10.1017/cbo9781139170215.007

Gould, S. J., *Time's Arrow, Time's Cycle: Myth and Metaphor in the Discovery of Geological Time* (Harmondsworth: Penguin, 1990).

Graham, A. C., 'China, Europe, and the Origins of Modern Science: Needham's The Grand Titration', in Nakayama, S. and Sivin, N. (eds.), *Chinese Science: Explorations of an Ancient Tradition* (Cambridge, MA and London: MIT Press, 1973), pp. 45-69.

Grant, E. (ed.), *A Source Book in Medieval Science* (Cambridge, MA: Harvard University Press, 1974).

— *The Foundations of Modern Science in the Middle Ages* (Cambridge: Cambridge University Press, 1996), http://dx.doi.org/10.1017/cbo9780511817908

Greenaway, F., 'Contribution to Discussion of Part Three: Science and Technology in the Middle Ages', in Crombie, A. C. (ed.), *Scientific Change* (London: Heinemann, 1963), pp. 329-31.

Hacking, I. (ed.), *Scientific Revolutions* (Oxford: Oxford University Press, 1981).

Hadden, R. W., *On the Shoulders of Merchants: Exchange and the Mathematical Conception of Nature in Early Modern Europe* (Albany, NY: State University of New York Press, 1994).

Hagberg, K., *Carl Linnaeus*, transl. Alan Blair (London: Cape, 1952).

Hahn, R., *The Anatomy of a Scientific Institution: The Paris Academy of Sciences, 1666-1803* (Berkeley, CA and London: University of California Press, 1971).

Hall, A. Rupert, 'Early Modern Technology to 1600', in Kranzberg, M. and Pursell, C. W., Jr. (eds.), *Technology in Western Civilization*, Vol. 1 (New York: Oxford University Press, 1967), pp. 79-103.

— *The Revolution in Science, 1500-1750* (London and New York: Longman, 1983), http://dx.doi.org/10.4324/9781315836850

Harkness, D. E., *The Jewel House: Elizabethan London and the Scientific Revolution* (New Haven, CT and London: Yale University Press, 2007).

Harnack, A., *Geschichte der königlich preussischen Akademie der Wissenschaften zu Berlin*, Vol. 2 (Urkunden u. Actenstücke) (Berlin: Reichsdruckerei, 1900).

Haubelt, J., 'František Josef Kinský', *Věstník Československé akademie věd*, 78 (1969), 560-77.

Heath, T. L., *The Works of Archimedes* (Cambridge: Cambridge University Press, 1912), available at https://archive.org/details/worksofarchimede029517mbp

Henry, J., *The Scientific Revolution and the Origins of Modern Science* (Basingstoke: Macmillan, 1997, 3rd ed. 2008).

Henshaw, Mr., 'The History Of the Making of Salt-Peter', in Sprat, T., *The History of the Royal Society in London* (London: Printed by T. R. for J. Martyn and J. Allestry, 1667) pp. 260-76.

[Henshaw, Mr.], 'The History Of Making Gunpowder', in Sprat, T., *The History of the Royal Society in London*, pp. 277-83.

Hessen, B., 'The Social and Economic Roots of Newton's "Principia" ', in *Science at the Cross Roads*, 2nd ed. (London: Cass, 1971), pp. 151-212. With a new Foreword by Needham, J. and a new Introduction by P. G. Werskey.

Hicks, J., *A Theory of Economic History* (repr. Oxford: Oxford University Press, 1973).

Hobsbawm, E., *On History* (London: Weidenfeld & Nicolson, 1997).

Holorenshaw, H., [J. Needham], 'The Making of an Honorary Taoist', in Teich, M. and Young, R. (eds.), *Changing Perspectives in the History of Science* (London: Heinemann Educational Books, 1973), pp. 1-20.

Hook, Mr., 'A Method For Making a History of the Weather', in Sprat, T., *The History of the Royal Society in London*, pp. 173-79.

[Hornigk, Ph. W.], *Oesterreich ueber alles wann es nur will*, 2nd edn ([n.p.]: [n. pub.], 1685).

Horský, Z., 'Die Wissenschaft am Hofe Rudolfs II', in *Prag um 1600 Kunst und Kultur am Hofe Kaiser Rudolf II* (Freren: Luca Verlag, 1988), pp. 69-74.

Hunter, M., *Boyle: Between God and Science* (New Haven, CT and London: Yale University Press, 2009).

Ihde, A. J., *The Development of Modern Chemistry* (New York: Evanston; London: Harper and Row, 1964).

Inkster, I., 'Thoughtful Doing and Early-Modern Oeconomy', in Roberts, L., Shaffer, S. and Dear, P. (eds.), *The Mindful Hand: Inquiry and Invention from the Late Renaissance to Early Industrialization* (Amsterdam: Koninkliijke Nederlandse Akademie van Wetenschappen, 2007), pp. 443-52

Isaac, G. L., 'Aspects of Human Evolution', in Bendall, D. S. (ed.), *Evolution from Molecules to Men* (Cambridge: Cambridge University Press, 1983), pp. 509-43.

Jacob, M. C., 'The Truth of Newton's Science and the Truth of Science's History: Heroic Science at its Eighteenth-Century Formulation', in M. J. Osler (ed.), *Rethinking the Scientific Revolution* (Cambridge: Cambridge University Press, 2000) pp. 315-32, http://dx.doi.org/10.1017/cbo9780511529276.016

Jardine, N., *The Birth of History and Philosophy of Science: Kepler's 'A Defence of Tycho against Ursus' with Essays on its Provenance and Significance* (Cambridge: Cambridge University Press, 1988).

Jardine, N. and Segonds, A.-Ph., *La Guerre des Astronomes: La Querelle au sujet de l'origine du système géo-héliocentrique à la fin du XVIe siècle*. Vol. 1: *Introduction* (Paris: Belles lettres, 2008).

— *La Guerre des Astronomes: La Querelle au sujet de l'origine du système géo-héliocentrique à la fin du XVIe siècle*. Vol. 2/1: *Le 'Contra Ursum de Jean Kepler'*, *Introduction et textes préparatoires* and Vol. 2/2: *Le 'Contra Ursum' de Jean Kepler*, *Édition critique, traduction et notes* (Paris: Belles lettres, 2008).

Johnson, Ch. (ed. and transl.), *The "De moneta" of Nicholas Oresme and English Mint Documents* (London: T. Nelson, 1956).

Jones, J., 'Natural Wonders', *Saturday Guardian*, 26 April 2008, http://www.theguardian.com/books/2008/apr/26/art.art.

Jones, S., Martin, R. and Pilbeam, D. (eds.), *The Cambridge Encyclopedia of Human Evolution* (Cambridge: Cambridge University Press, 1992).

Kaufmann DaCosta, T., *Arcimboldo: Visual Jokes, Natural History, and Still-Life Painting* (Chicago, IL: University of Chicago Press, 2009).

Kaye, J., *Economy and Nature in the Fourteenth Century: Money, Market Exchange, and the Emergence of Scientific Thought* (Cambridge: Cambridge University Press, 1998), http://dx.doi.org/10.1017/cbo9780511496523

Kiernan, V. G., *State and Society in Europe, 1550-1650* (Oxford: Blackwell, 1980).

Kinsky, Fr., 'Schreiben des Herrn Grafen von K... an Herrn von Born ueber einige mineralogische und lithologische Merkwuerdigkeiten', *Abhandlungen*, 1 (1775), 243-52.

Knight, D. M., *Voyaging in Strange Seas: The Great Revolution in Science* (New Haven, CT and London: Yale University Press, 2014).

Koerner, L., *Linnaeus: Nature and Nation* (Cambridge, MA: Harvard University Press, 1999).

Koyré, A., *Études galiléennes*, 3 vols. (Paris: Hermann, 1939-1940).

Krohn, W., 'Zur Geschichte des Gesetzesbegriffs in Naturphilosophie und Naturwissenschaft', in Hahn, M. und Sandkühler, H.-J. (eds.), *Gesellschaftliche Bewegung und Naturprozess* (Cologne: Pahl-Rugenstein, 1981), pp. 61-70.

— *Francis Bacon* (Munich: C. H. Beck, 1987).

— 'Social Change and Epistemic Thought (Reflections on the Origin of the Experimental Method)', in Hronszky, I., Fehér, M., and Dajka, B. (eds.), *Scientific Knowledge Socialized*, Boston Studies in the Philosophy of Science, Vol. 108 (Dordrecht, Boston and London: Kluwer, 1988), pp. 165-78.

Kuhn, T. S., 'The Function of Dogma in Scientific Research', in Crombie, A. C. (ed.), *Scientific Change* (London: Heinemann, 1963), pp. 347-69.

— *The Structure of Scientific Revolutions*, 2nd revised ed. (Chicago, IL: University of Chicago Press, 1970).

Landes, D. S., *The Unbound Prometheus: Technological Change and Industrial Development in Western Europe from 1750 to the Present* (Cambridge: Canbridge University Press, 1969).

— *Revolution in Time* (Cambridge, MA: Harvard University Press, 1985).

Leakey, R. E., *The Origin of Humankind* (London: Basic Books, 1994).

Leibniz, 'Errichtung einer Societät in Deutschland (2. Entwurf)', in Harnack, A., *Geschichte der königlich preussischen Akademie der Wissenschaften zu Berlin*, Vol. 2 (Urkunden u. Actenstücke) (Berlin: Reichsdruckerei, 1900), pp. 19-26.

Lindberg, D. C. and Westman, R. S. (eds.), *Reappraisals of the Scientific Revolution* (Cambridge: Cambridge University Press, 1990).

Lindroth, S., 'The Two Faces of Linnaeus', in Frängsmyr, T. (ed.), *Linnaeus: The Man and his Work* (Berkeley, Los Angeles, CA and London: University of California Press, 1983), pp. 1-62.

Lloyd, G. E. R., *Early Greek Science: Thales to Aristotle* (London: Chatto & Windus, 1970).

— *Greek Science after Aristotle* (London: Chatto and Windus, 1973).

— *Methods and Problems in Greek Science: Selected Papers* (Cambridge: Cambridge University Press, 1991).

— 'Greek Antiquity: The Invention of Nature', in Torrance, G. (ed.), *The Concept of Nature: The Herbert Spencer Lectures* (Oxford: Clarendon Press, 1992). Reprinted as 'The Invention of Nature', in Lloyd, G. E. R., *Methods and Problems in Greek Science* (Cambridge: Cambridge University Press, 1991), pp. 417-34.

— 'Democracy, Philosophy and Science in Ancient Greece', in J. Dunn (ed,. *Democracy: The Unfinished Journey, 508 BC to AD 1993* (Oxford: Oxford University Press, 1993), pp. 41-56.

— *Adversaries and Authorities: Investigations into Ancient and Greek Chinese Science* (Cambridge: Cambridge University Press, 1996).

Machamer, P. (ed.), *The Cambridge Companion to Galileo* (Cambridge: Cambridge University Press, 1998), http://dx.doi.org/10.1017/ccol0521581788

Marx, K., *Capital*, Vol. 1 (London: George Allen & Unwin, 1938).

Mason, P., 'What Shakespeare Taught Me about Marxism and the Modern World', *The Guardian*, 3 November, 2013, http://www.theguardian.com/commentisfree/2014/nov/02/sharkespeare-marxism-feudalism-capitalism.

Mayer, H., 'Gott und Mechanik Anmerkung zur Geschichte des Naturbegriffs im 17. Jahrhundert', in Mattl, S. and others, *Barocke Natur* (Korneuburg: Ueberreuter, 1989), pp. 12-25.

Metzger, H.-D., *Thomas Hobbes und die Englische Revolution 1640-1660* (Stuttgart-Bad Constatt: Frommann-Holzboog, 1991).

Mirsky, J. 'Tribute, Trade and Some Eunuchs', *The Times Literary Supplement*, 26 January 2007, 11.

Mitterauer, M., *Why Europe? The Medieval Origin of its Special Path* (Chicago, IL and London: University of Chicago Press, 2010), http://dx.doi.org/10.7208/chicago/9780226532387.001.0001

Mokyr, J., *The Gifts of Athena: Historical Origins of the Knowledge Economy* (Princeton, NJ and Oxford: Princeton University Press, 2005).

Musson, A. E. and Robinson, E., *Science and Technology in the Industrial Revolution* (Manchester: Manchester University Press, 1969).

Needham, J., 'On Science and Social Change', *Science and Society*, 10 (1946), 225-51.

— *Science and Civilisation in China*, Vols. 2-3 (Cambridge: Cambridge University Press, 1956-1959).

— 'Science and Society in East and West', in Goldsmith, M. and Mackay, A. (eds.), *The Science of Science: Society in Technological Age* (London: Souvenir Press, 1964), pp. 127-49.

— *The Grand Titration: Science and Society in East and West* (London: Allen & Unwin, 1969).

— 'Abstract of Material Presented to the International Maritime History Commission at Beirut', in Mollat, M. (ed.), *Sociétés et Compagnies de Commerce en Orient et dans L'Océan Indien* (Paris: S.E.V.P.E.N, 1970), pp. 139-65.

— *Moulds of Understanding: A Pattern of Natural Philosophy*, ed. and intr. by G. Werskey (London: Allen and Unwin, 1976).

— *Wissenschaftlicher Universalismus Über Bedeutung und Besonderheit der chinesischen Wissenschaft*, ed., intr. and transl. by T. Spengler (Frankfurt am Main: Suhrkamp, 1979).

Needham, J. with Wang Ling, *Science and Civilisation in China*, Vol. 4/2 (Cambridge: Cambridge University Press, 1965).

Needham, J. with Wang Ling and Lu Gwei-Djen, *Science and Civilisation in China*, Vol. 4/3: 'Civil Engineering and Nautics' (Cambridge: Cambridge University Press, 1971).

Needham, J. and Huang, R. (Huang Jen-Yü), 'The Nature of Chinese Society – A Technical Interpretation', *Journal of Oriental Studies*, repr. from Vol. 12/1 and 2 (1974), 1-16.

North, J., *Cosmos: An Illustrated History of Astronomy* (Chicago, IL and London: University of Chicago Press, 2008).

O'Brien, P. K., 'The Needham Question Updated: A Historiographical Survey and Elaboration', *History of Technology*, 29 (2009), pp. 7-28.

Ogilvie, S. C. and Cerman, M. (eds.), *European Proto-industrialization: An Introductory Handbook* (Cambridge: Cambridge University Press, 1996).

Oresme, N., *The Configurations of Qualities and Motions, including a Geometric Proof of the Mean Speed Theorem*, trans., intr. and annotated by Marshall Clagett in Grant, E. (ed.), *A Source Book in Medieval Science* (Cambridge, MA: Harvard University Press, 1974).

Osler, M. J. (ed.), *Rethinking the Scientific Revolution* (Cambridge: Cambridge University Press, 2000), http://dx.doi.org/10.1017/cbo9780511529276

— 'The Canonical Imperative: Rethinking the Scientific Revolution', in Osler, M., *Rethinking the Scientific Revolution*, pp. 3-22.

Oster, M. (ed.), *Science in Europe, 1500-1800: A Primary Sources Reader* (Basingstoke: Palgrave, 2002).

Pagel, W., *William Harvey's Biological Ideas: Selected Aspects and Historical Background* (Basel and New York: Karger, 1967).

Pedersen, O. and Pihl, M., *Early Physics and Astronomy: A Historical Introduction* (London: Macdonald and Jane's and New York: American Elsevier Inc, 1974).

Petráň, J. and Petráňová, L., 'The White Mountain as a Symbol in Modern Czech History', in Teich, M. (ed.), *Bohemia in History* (Cambridge: Cambridge University Press, 1998), pp. 143-63.

Petty, Sir William, 'An Apparatus to the History of the Common Practices of Dying', in Sprat, T., *The History of the Royal Society in London*, pp. 284-306.

Pickstone, J. V., *Ways of Knowing: A New History of Science, Technology and Medicine* (Manchester: Manchester University Press, 2000).

Poda, N. von, *Kurzgefasste Beschreibung der, bey dem Bergbau zu Schemnitz in Nieder-Hungarn, errichteten Maschinen etc.* (Prague: Walther, 1771).

Polišenský, J. V. with Snider, F., *War and Society in Europe 1618-1648* (Cambridge: Cambridge University Press, 1978), http://dx.doi.org/10.1017/cbo9780511897016

Polišenský, J., *Komenský Muž labyrintů a naděje* [*Komenský Man of Labyrinths and Hope*] (Prague: Academia, 1996).

Pollock, S., *Ancient Mesopotamia: The Eden that Never Was* (Cambridge: Cambridge University Press, 1999).

Porter, R., 'The Scientific Revolution: A Spoke in The Wheel?', in Porter, R. and Teich, M. (eds.), *Revolution in History* (Cambridge: Cambridge University Press, 1986), pp. 290-316.

Porter, R., and Teich, M., 'Introduction', in Porter and Teich (eds.), *The Scientific Revolution*, pp. 1-10, http://dx.doi.org/10.1017/cbo9781139170215.001

Porter, R. and Teich, M. (eds.), *The Enlightenment in National Context* (Cambridge: Cambridge University Press, 1981), http://dx.doi.org/10.1017/cbo9780511561283

— *Revolution in History* (Cambridge: Cambridge University Press, 1986).

— *Romanticism in National Context* (Cambridge: Cambridge University Press, 1988).

— *The Renaissance in National Context* (Cambridge: Cambridge University Press, 1991).

— *The Scientific Revolution in National Context* (Cambridge: Cambridge University Press, 1992), http://dx.doi.org/10.1017/cbo9781139170215

Pumfrey, S., '"Your Astronomers and Ours Differ Exceedingly": The Controversy over the "New Star" of 1572 in the Light of a Newly Discovered Text by Thomas Digges', *The British Journal for the History of Science*, 44 (2011), 29-60, http://dx.doi.org/10.1017/s0007087410001317

Purš, I. and Karpenko, V. (eds.), *Alchymie a Rudolf II: Hledání tajemství přírody ve střední Evropě v 16. a 17. století* [*Alchemy and Rudolf II Searching for Secrets of Nature in Central Europe in the 16th and 17th Century*] (Prague: Artefactum Ústav dějin umění AV ČR, 2011).

Reinalter, H. (ed.), *Die Aufklärung in Österreich. Ignaz von Born und seine Zeit* (Frankfurt am Main; Bern; New York; Paris: Lang, 1991).

Ritter, J., 'Metrology, Writing and Mathematics in Mesopotamia', *Acta historiae rerum naturalium necnon technicarum. Prague Studies in the History of Science and Technology*, N. S. (1999), 215-41.

Rose, S., 'Mathematics and the Art of Navigation: The Advance of Scientific Seamanship in Elizabethan England', *Transactions of the Royal Historical Society*, 14 (2004), 175-84, http://dx.doi.org/10.1017/S0080440104000192

Rossi, P., *Francis Bacon: From Magic to Science* (Chicago, IL: University of Chicago Press, 1968).

Rudolph, H., 'Kirchengeschichtliche Beobachtungen zu J. J. Becher', in Frühsorge, G. and Strasser, G. F. (eds.), *Johann Joachim Becher (1635-1682)* (Wiesbaden: Harrassowitz, 1993), pp. 173-96.

Schaffer, S., 'Golden Means: Assay Instruments and the Geography of Precision in the Guinea Trade', in Bourget, M.-N., Licoppe, Ch. and Sibum, H. O. (eds.), *Instruments, Travel and Science: Itineraries of Precision from the Seventeenth to the Twentieth Century* (London and New York: Routledge, 2002), pp. 20-50.

Scribner, B., Porter, R. and Teich, M. (eds.), *The Reformation in National Context* (Cambridge: Cambridge University Press, 1994), http://dx.doi.org/10.1017/CBO9780511599569

Shapin, S., 'Discipline and Bounding: The History and Sociology of Science Seen through the Externalism-Internalism Debate', *History of Science*, 30 (1992), 333-69, available at http://dash.harvard.edu/handle/1/3425943

— *The Scientific Revolution* (Chicago, IL and London: University of Chicago Press, 1998).

Shapin, S. and Schaffer, S., *Leviathan and the Air-Pump: Hobbes, Boyle, and the Experimental Life* (Princeton, NJ: Princeton University Press, 1985).

Sivin, N., 'An Introductory Bibliography of Traditional Chinese Science. Books and Articles in Western Languages', in Nakayama and Sivin, *Chinese Science Explorations*, pp. 279-314.

— 'Why the Scientific Revolution Did Not Take Place in China – or Didn't It?', in Mendelsohn, E. (ed.), *Transformation and Tradition in the Sciences: Essays in Honor of I. Bernard Cohen* (Cambridge: Cambridge University Press, 1984), pp. 531-54.

Skinner, Q., *Hobbes and Republican Liberty* (Cambridge: Cambridge University Press, 2008).

Smith, P. H., *The Business of Alchemy: Science and Culture in the Holy Roman Empire* (Princeton, NJ: Princeton University Press, 1994).

Smolka, J., 'Scientific Revolution in Bohemia', in Porter and Teich, *The Scientific Revolution*, pp. 210-39, http://dx.doi.org/10.1017/cbo9781139170215.009

— 'Böhmen und die Annahme der Galileischen astronomischen Entdeckungen', *Acta historiae rerum naturalium necnon technicarum. Prague Studies in the History of Science and Technology*, Vol. 1 (1997), pp. 41-69.

— 'Thaddaeus Hagecius ab Hayck, Aulae Caesarae Majestatis Medicus', in Enderle-Burcel, G., Kubů, E., Šouša, J. and Stiefel, D. (eds.), *"Discourses – Diskurse" Essays for – Beiträge zu Mikuláš Teich & Alice Teichova* (Prague and Vienna: Nová tiskárna Pelhřimov, 2008), pp. 395-412.

Sprat, T., *The History of the Royal Society in London* (London: Printed by T. R. for J. Martyn and J. Allestry, 1667). Repr. and ed. with critical apparatus by Cope, J. I. and Jones, H. W. (St. Louis, MI: Washington University Studies, 1958).

Stern, L., *Zur Geschichte und wissenschaftlichen Leistung der Deutschen Akademie der Naturforscher "Leopoldina"* (Berlin: Rütten & Loening, 1952).

Stewart, M., *The Courtier and the Heretic: Leibniz and Spinoza and the Fate of God in the Modern World* (New Haven, CT and London: Yale University Press, 2005).

Strawson, G., 'Descartes and Elisabeth', *The Times Literary Supplement*, 13 February, 2015, 6.

Švejda, A., 'Prager Konstrukteure wissenschaftlicher Instrumente und ihre Werke', in Folta, J. (ed.), 'Science and Technology in Rudolfinian Time', *Acta historiae rerum naturalium necnon technicarum Prague Studies in the History of Science and Technology*, Vol. 1 (1997), pp. 90-4.

Teich, M., 'Tschirnhaus und der Akademiegedanke', in Winter, E. (ed.), *E. W. von Tschirnhaus und die Frühaufklärung in Mittel- und Osteuropa* (Berlin: Akademie Verlag, 1960), pp. 93-107.

— 'The Two Cultures, Comenius and the Royal Society', *Paedagogica Evropea*, 4 (1968), 147-53.

— 'Afterword', in Porter, R. and Teich, M. (eds.), *The Enlightenment in National Context* (Cambridge: Cambridge University Press, 1981), pp. 215-17, http://dx.doi.org/10.1017/cbo9780511561283.015

— 'Bohemia: From Darkness into Light', in Porter, R. and Teich, M. (eds.), *The Enlightenment in National Context* (Cambridge: Cambridge University Press, 1981), pp. 141-63, http://dx.doi.org/10.1017/cbo9780511561283.011

— 'Circulation, Transformation, Conservation of Matter and the Balancing of the Biological World in the Eighteenth Century', *Ambix*, 29 (1982), 17-28.

— 'The Scientific-Technical Revolution: An Historical Event in the Twentieth Century?', in Porter and Teich, *Revolution in History*, pp. 317-30.

— 'Interdisciplinarity in J. J. Becher's Thought', in Frühsorge, G. and Strasser, G. F. (eds.), *Johann Joachim Becher (1635-1682)* (Wiesbaden: Harrassowitz, 1993), pp. 23-40 (previously published in *History of European Ideas*, 9 (1988), 145-60, http://dx.doi.org/10.1016/0191-6599(88)90036-8

— 'Revolution, wissenschaftliche', in Sandkühler, H. J. (ed.), *Enzyklopädie Philosophie*, Vol. 2. O-Z (Hamburg: Meiner, 1999), pp. 1394-97.

— 'J. D. Bernal: The Historian and the Scientific Technical Revolution', *Interdisciplinary Science Reviews*, 33 (2008), 135-39, http://dx.doi.org/10.1179/030801808x259790

Teich, M. (ed.), *Bohemia in History* (Cambridge: Cambridge University Press, 1998).

Teich, M. with Dorothy M. Needham, *A Documentary History of Biochemistry, 1770-1940* (Leicester and London: Leicester University Press, 1992).

Teich, M. and Porter, R. (eds.), *The Industrial Revolution in National Context: Europe and the USA* (Cambridge: Cambridge University Press, 1996).

Teich, M., Porter, R. and Gustafsson, B. (eds.), *Nature and Society in Historical Context* (Cambridge: Cambridge University Press, 1997).

Teich, M. and Young, R. (eds.), *Changing Perspectives in the History of Science* (London: Heinemann Educational Books, 1973).

Teich, N., 'Influence of Newton's Work on Scientific Thought', *Nature*, 153 (1944), 42-5, http://dx.doi.org/10.1038/153042a0

Válka, J., 'Rudolfine Culture', in Teich, M. (ed.), *Bohemia in History* (Cambridge: Cambridge University Press, 1998), pp. 117-42.

Van der Wee, H., 'The Influence of Banking on the Rise of Capitalism in North-West Europe, Fourteenth to Nineteenth Century', in Teichova, A., Kurgan-van Hentenryk, G. and Ziegler, D. (eds.), *Banking, Trade and Industry: Europe, America and Asia from the Thirteenth to the Twentieth Century* (Cambridge: Cambridge University Press, 1997), pp. 173-88.

Vandermeer, J., 'The Agroecosystem: The Modern Vision Crisis, The Alternative Evolving', in Singh, R., Krimbas, C. B., Paul, D. B. and Beatty, J. (eds.), *Thinking about Evolution: Historical, Philosophical and Political Perspectives*, Vol. 2 (Cambridge: Cambridge University Press, 2001), pp. 480-509.

Waerden, B. L. van der, 'Basic Ideas and Methods of Babylonian and Greek Astronomy', in Crombie, A. C. (ed.), *Scientific Change* (London: Heinemann, 1963), pp. 42-60.

Webster, C. (ed.), *Samuel Hartlib and the Advancement of Learning* (Cambridge: Cambridge University Press, 1970).

— *The Great Instauration: Science, Medicine and Reform, 1626-1660* (London: Duckworth, 1975).

Weld, C. R., *A History of the Royal Society*, 2 vols. (London: J. W. Parker, 1848).

Westfall, R. S., 'The Scientific Revolution Reasserted', in Osler, M. J. (ed.), *Rethinking the Scientific Revolution* (Cambridge: Cambridge University Press, 2000), pp. 41-55, http://dx.doi.org/10.1017/cbo9780511529276.004

Whaley, J., *Germany and the Holy Roman Empire*, 2 vols. (Oxford: Oxford University Press, 2012).

Wilson, P. H., *Europe's Tragedy: A History of the Thirty Years War* (London: Penguin, 2010).

Winter, E., 'Die katholischen Orden und die Wissenschaftspolitik im 18. Jahrhundert', in Amburger, E., Cieśla, M. C. and Sziklay, L. (eds.), *Wissenschaftspolitik in Mittel- und Osteuropa* (Berlin: Camen, 1976), pp. 85-96.

Wolff, M., *Geschichte der Impetustheorie* (Frankfurt am Main: Suhrkamp, 1978).

— 'Mehrwert und Impetus bei Petrus Johannis Olivi', in Miethke, J. and Schreiner, K. (eds.), *Sozialer Wandel im Mittelalter* (Sigmaringen: Thorbecke, 1994), pp. 413-23.

Wondrák, E., 'Die Olmützer "Societas incognitorum". Zum 225. Jubiläum ihrer Gründung und zum 200. Todestag ihres Gründers', in Lesky, E., Kostić, D.S.K., Matl, J. and Rauch, G. v. (eds.), *Die Aufklärung in Ost- und Südosteuropa* (Cologne and Vienna: Böhlau, 1972), pp. 215-28.

Yabuuti, K., 'Chinese Astronomy: Development and Limiting Factors', in Nakayama and Sivin, *Chinese Science Explorations*, pp. 91-103.

Zilsel, E., 'The Sociological Roots of Science' [1942], repr. in Raven, D., Krohn, W. and Cohen, R. S. (eds.), *Edgar Zilsel: The Social Origins of Modern Science* (Dordrecht , Boston, MA and London: Kluwer, 2000), pp. 7-21.

Index

This book need not end here...

At Open Book Publishers, we are changing the nature of the traditional academic book. The title you have just read will not be left on a library shelf, but will be accessed online by hundreds of readers each month across the globe. We make all our books free to read online so that students, researchers and members of the public who can't afford a printed edition can still have access to the same ideas as you.

Our digital publishing model also allows us to produce online supplementary material, including extra chapters, reviews, links and other digital resources. Find The *Scientific Revolution Revisited* on our website to access its online extras. Please check this page regularly for ongoing updates, and join the conversation by leaving your own comments:

http://www.openbookpublishers.com/isbn/9781783741229

If you enjoyed this book, and feel that research like this should be available to all readers, regardless of their income, please think about donating to us. Our company is run entirely by academics, and our publishing decisions are based on intellectual merit and public value rather than on commercial viability. We do not operate for profit and all donations, as with all other revenue we generate, will be used to finance new Open Access publications.

For further information about what we do, how to donate to OBP, additional digital material related to our titles or to order our books, please visit our website: http://www.openbookpublishers.com

OpenBook Publishers

Knowledge is for sharing

Lightning Source UK Ltd.
Milton Keynes UK
UKOW06f2304140415

249341UK00002B/1/P